Infrastructure Investments in Developing Economies

Giang Dang · Low Sui Pheng

Infrastructure Investments in Developing Economies

The Case of Vietnam

Springer

Giang Dang
Low Sui Pheng
Department of Building
National University of Singapore
Singapore
Singapore

ISBN 978-981-10-1267-9 ISBN 978-981-287-248-7 (eBook)
DOI 10.1007/978-981-287-248-7

Springer Singapore Heidelberg New York Dordrecht London

Printed on acid-free paper

Springer is part of Springer Science+Business Media (www.springer.com)

Preface

This book studies critical factors affecting the efficient use of public investments in infrastructure to support trade and economic development. In striving for that goal, the study reviewed the literature on construction and economic development, international trade theories, and infrastructure development. The literature review provided an understanding of the linkages between the economy, infrastructure, and trade growth. The literature has shown a positive relationship between trade and economic development. Stimulating trade has been one of the major policies in developing countries. However, the underperformance of the infrastructure systems that links these countries with their international markets is considered one of the major concerns. Investments in infrastructure have increasingly been used to improve the infrastructure quality, thus stimulating trade and economic growth in developing countries. Nevertheless, a number of infrastructure challenges faced by developing countries remain. This raises the question about the efficiency in the use of public investments in infrastructure.

Examining how to use public investments in infrastructure more efficiently to support trade and economic development, this study found the answer in the way infrastructure of a country is planned and built. This study also attempts to isolate key factors in the process that could limit the efficient use of public investments in infrastructure. From the literature review, these factors could be categorized into two groups: factors in the planning process and factors in the implementation process. In the planning process, these factors include inadequate capacity for estimating and monitoring of rates of return of infrastructure projects; politicized decision making; transparency and accountability problems; and institutional weaknesses in decision making. In the implementation process, these factors include lack of political commitment in the implementation of infrastructure development plans; problems in land acquisition; corruption in infrastructure construction; shortages of local construction firms' capabilities and resources; and institutional and legal weaknesses in infrastructure construction.

This book discusses these factors in the case of Vietnam. A longitudinal survey was carried out to examine the developmental trends of the efficiency of public investments in infrastructure construction to support trade and economic growth over time. A questionnaire survey and in-depth interviews with respondents from both government agencies and nongovernment agencies were conducted to identify key factors affecting the efficient use of the public investments. Respondents from the government sector were selected from relevant government agencies involved in policy making as well as planning for construction, infrastructure, and trade of Vietnam. Respondents from nongovernment agencies included financial institutions, traders, developers, consultants, and contractors involved in investing and delivering infrastructure, as well as in utilizing infrastructure services for trading activities. There are four major factors that this book found critical to the current use of public investments in infrastructure in Vietnam including institutional capacity of the government, transparency and accountability in infrastructure development, land acquisition, and building capacity of local firms. A model of the impacts of these factors on the efficient use of public investments in infrastructure was examined and validated. Among these factors, poor institutional capacity of the government lies at the root of the inefficient use of public investments in infrastructure. The current fragmented way of planning and implementing infrastructure policies requires better coordination across sectors and levels.

Findings of the study supplement the understanding of the role of government in the theory of coordination failure in economic development. Major theoretical implication of the findings is that there is a strong need for the government to take the lead in ensuring that public investments in infrastructure lead to long-term and sustainable growth. Practical contributions of the study are also achieved by more precisely pinning down constraints on state capacities across infrastructure planning and delivery, explaining the effects of these constraints, and illuminating the potential scope for reform in policy and planning for infrastructure development.

Contents

Contents

Abbreviations

ADB	Asian Development Bank
APEC	Asia–Pacific Economic Cooperation
ASEAN	Association of Southeast Asian Nations
AVE	Average Variance Extracted
CB-SEM	Covariance-based SEM
ECE	United Nations Economic Commission for Europe
EVN	Electricity of Vietnam
FDI	Foreign Direct Investment
FIEs	Foreign-Invested Enterprises
GATS	General Agreement for Trade in Services
GATT	General Agreement on Tariffs and Trade
GCI	Global Competitiveness Index
GCR	Global Competitiveness Report
GDP	Gross Domestic Product
GFCF	Gross Fixed Capital Formation
GNI	Gross National Income
GNP	Gross National Product
GSO	General Statistics Office of Vietnam
HĐBT	Ministers' Committee
HDI	Human Development Index
MDGs	Millennium Development Goals
MoC	Ministry of Construction
MoF	Ministry of Finance
MoIC	Ministry of Information and Communications
MoIT	Ministry of Industry and Trade
MoT	Ministry of Transport
MPI	Ministry of Planning and Investment
MSA	Measure of Sampling Adequacy
NICs	Newly Industrializing Countries
ODA	Official Development Assistance

PIP	Public Investment Program
PLS-SEM	Partial Least Squares SEM
PMU	Project Management Unit
PNTR	Permanent Normal Trade Relations
PPP	Public–Private Partnership
SEDP	Socio-Economic Development Plan
SEDS	Socio-Economic Development Strategy
SEM	Structural Equation Modeling
SOEs	State-Owned Enterprises
SPS	Sanitary and Phytosanitary Standards
SPSS	Statistical Package for Social Sciences
TEC & QMB	Transport Engineering Construction and Quality Management Bureau
TRIPS	Trade Related Intellectual Property Rights
UNIDO	United Nations Industrial Development Organization
UNSD	United Nations Statistics Division
VACC	Vietnam Association of Construction Contractors
VBRA	Vietnam Bridge and Road Association
VEC	Vietnam Expressway Corporation
VFCEA	Vietnam Federation of Civil Engineering Associations
VNPT	Vietnam Posts and Telecommunications Group
VNR	Vietnam Railways Corporation
VUSTA	Vietnam Union of Science and Technology Associations
WEF	World Economic Forum
WTO	World Trade Organization

Chapter 1
Introduction

Abstract This chapter introduces the subject matters of this study, which is to examine the question of efficiency in using public investments in infrastructure to support trade and economic growth in developing countries in general and in the context of Vietnam in particular. This chapter also summarizes the major research methodology used in the study and presents the structure of the book.

Keywords Infrastructure · Trade · Economic development · Developing countries · Vietnam

1.1 Research Problem

International trade today stands as a focus of the development agenda in many countries of the world. Notably, international trade in developing countries has recently grown faster than in developed ones (World Trade Organization 2010). In 2010, the growth rates of exports in developed countries and developing countries were 13 and 17 %, respectively. Likewise, the 2010 growth rate of imports in developing countries (18 %) was higher than that in developed countries (11 %) (World Trade Organization 2010). However, integrating into the world trade system is still a challenging task for developing countries. Traditional trade issues are barriers to trade (tariffs and quotas), which remain high in many developing countries. As developing countries increase their integration into the world economy, these countries are facing new issues. They are "behind-the-borders" issues, including the role of infrastructure and governance (Hoekman et al. 2002).

The role of infrastructure in supporting trade as well as infrastructure issues in the developing countries has been addressed in a number of empirical studies. In these studies, infrastructure is regarded as one of the determinants of international competitiveness, which directly impact the ability of countries to engage in international trade, and to compete for foreign direct investments (APEC Economic Committee 1997; Nwankwo 2000; Kohsaka 2007). The poor quality and inadequacy of

G. Dang and L. Sui Pheng, *Infrastructure Investments in Developing Economies*,
DOI 10.1007/978-981-287-248-7_1

infrastructure are therefore considered one of the major marketing issues and challenges in many developing countries (Batra 1997). As a result, attention has increasingly been given to strategic infrastructure development in the developing countries. Most studies on strategic infrastructure development for developing countries examined issues related to financing issues for infrastructure development, including under-investment in infrastructure, allocations of public capital, public resource constraints and private financing (Mundial 1994; Venables 2004; Demetriades 2007; Kohsaka 2007; Flanagan and Jewell 2009).

Similarly, the strong correlation between construction (including infrastructure) and economic development in the developing countries has long been studied (Turin 1969; World Bank 1984). Tan (2002) has summarized the implications of the early studies of the relation between construction and economic development in the less developed countries and noted that construction is a "potential" agent of economic development.

There are, however, other concerns about the infrastructure development in developing countries, including waste, cost overruns and benefit shortfalls, or corruption in large infrastructure projects (Flyvbjerg 2007; Sohail and Cavill 2008). These concerns suggest that there are more major management issues with infrastructure development than just project financing issues. Several studies focus on the planning aspect of infrastructure development (Hossain 2000; Flyvbjerg 2007; Mustajab 2009; Marshall 2011). However, little is known about how the entire management system (including planning and delivery) of the government can use public investments in infrastructure more efficiently to support trade and economic development. To this end, it is pertinent to examine how infrastructure is planned and built in a typical developing country, and what key variables are there in the system limiting the efficient use of public investments in infrastructure.

Vietnam has shown constant efforts and commitments to stimulate the integration into the world trading systems and heavy investments in infrastructure for economic development over the past decades (see Chap. 6). However, compared to other regional countries such as Singapore, Malaysia, Thailand, Indonesia and the Philippines, Vietnam's GNI per capita is the lowest (World Bank 2010). Similarly, there is a large gap in infrastructure between Vietnam and these neighbouring countries. According to the World Economic Forum (Schwab 2008, 2009, 2010), infrastructure quality in Vietnam was ranked the poorest compared to that of Singapore, Malaysia, Thailand, Indonesia and the Philippines (more details in Chap. 6). The weaknesses of infrastructure in Vietnam have been identified as the biggest hindrance for doing business in the country (Nguyen and Dapice 2009). It is therefore critical for Vietnam to develop plans and policies to overcome these challenges. As the country continues to rely on government planning to foster economic development, which is common in developing economies (Todaro and Smith 2009), the case of Vietnam could bring out key insights that go beyond the case under study.

1.2 Research Scope

In examining the main research question of what factors limit the efficient use of public investments in infrastructure to support trade and economic development in Vietnam, the approach taken in this study has several features:

- The research focuses on macrolevel or statewide variables rather than micro- or agency-level variables. The approach is needed since public investments in infrastructure (national infrastructure investment and policy) are initially determined at macrolevel in response to or stimulate changes in the economic environment.
- Trade in this study is referred to as trade between nations or international trade, since this study focuses on a country's perspective on trade rather than on an individual's or firm's perspective.
- There are two main types of infrastructure, economic infrastructure supporting economic activities of the national economy and social infrastructure supporting social development. Since the study focuses on infrastructure built to support economic development and trade, the scope of this study is restricted to economic infrastructure [as defined by World Bank (1994)], including power, telecommunications and transportation.
- Infrastructure construction in this study is referred to as the production process of the built environment, including various activities from conception through design to execution that transforms various resources into constructed facilities (Ive and Gruneberg 2000).
- In economics, efficiency is defined as the use of resources for maximizing the production of goods and services (O'Sullivan and Sheffrin 2004). In this study, the efficient use of public investments in infrastructure refers to the use of public investments in infrastructure so as to maximize the competitiveness of a nation in terms of infrastructure development. According to the World Economic Forum, the competitiveness of a nation in terms of infrastructure development is reflected in the quality and extensiveness of infrastructure networks that integrate the national market and connecting it at low cost to markets in other countries and regions (Schwab 2009).
- Developing countries in this study are referred to as low-income and middle-income economies as classified by the World Bank (2011). The other group includes developed countries, which have high incomes and high-living standards (World Bank 2011).

1.3 Research Aim and Objectives

The main aim of this study was to analyse the use of public investments in infrastructure to support trade and economic development in Vietnam. The objectives of this study are as follows:

- To review how public investments in infrastructure are used to support trade and economic growth in the literature;
- To examine how infrastructure is planned and built in Vietnam;
- To evaluate the level of efficiency of public investments in infrastructure in supporting trade and economic growth in Vietnam;
- To explore the key variables in the process of planning and plan implementation that limit the efficient use of public investments in infrastructure in Vietnam; and
- To make recommendations on capacity building to address issues that critically limit the efficient use of public investments in infrastructure to support trade and economic growth in Vietnam.

1.4 Research Methodology

The research methodology for this study is discussed in Chap. 7. However, it is useful to identify some features of the methodology at the beginning of this study:

- A longitudinal survey of public investments in infrastructure was used to explore the extent to which public investments in infrastructure was efficiently used in Vietnam;
- A questionnaire survey was adopted in this study to explore the key variables limiting the efficient use of public investments in infrastructure perceived by concerned parties. The sample includes government officials, developers, bankers, traders, consultants and contractors involved in policy-making and planning, and delivering as well as in utilizing infrastructure. A five-point Likert scale was applied for questions relating to the degree of occurrence and influence of factors;
- In-depth interviews were carried out to explore the ways to address key factors identified from the questionnaire survey. In addition, strategic planning documents, programme-related documents including final reports, evaluation reports and other documents were collected as alternative sources for cross-checking. Information collected from the interviews formed the main basis for the study to make recommendations to improve the efficient use of public investments in infrastructure.
- Data analysis included both quantitative (factor analysis and structural equation modelling) and qualitative (thematic content analysis) methods as explained in Chap. 7.
- A two-stage (quantitative and qualitative) validation was conducted to validate the results.

1.5 Research Significance

Economists have long agreed that the productive capacity of an economy depends on an adequate level of capital formation (Harrod 1948; Rostow 1960). Investments in infrastructure are expected to promote trade and other economic activities. Considerable financing is required as demand for infrastructure in developing countries to support trade, and economic growth is increasing rapidly. The question of project financing is thus the major subject discussed in the literature on infrastructure development (Fay and Yepes 2003; Davis 2008; Arnold 2011). However, as important as the question of project financing, developing countries are facing the question of efficiency. The poor performance of infrastructure investments (cost overruns, benefit shortfalls and waste) can have negative impacts on the economy and be costly to fix (Burns and Grebler 1984; Devarajan et al. 1996; Flyvbjerg 2008). Therefore, how efficiently public investments in infrastructure can be used should be considered. More importantly, what factors are limiting the efficient use of public investments in infrastructure should be explored. The research is thus needed to provide knowledge on how to better use public investments in infrastructure to benefit the economy.

The research explored the use of public investments in infrastructure at the macrolevel in order to identify constraints across government departments and policy spheres. Although development economists have long been debating the role of markets over government (Meier 2000), infrastructure investment is provided principally by the public sector worldwide (Estache 2006). Besides providing infrastructure to promote private economic activities, governments also often use infrastructure spending as a counter-cycle tool for macroeconomic stabilization (Gruneberg 1997; Hillebrandt 2000). Therefore, examining constraints across government departments and policy spheres, thus assisting government to put in place required capacities to improve the public sector performance is critical for developing countries. These insights are also valuable for policy-makers to develop better infrastructure policies when financial resources are tight.

In addition, the study explored the use of public investments in infrastructure by employing a macrolevel approach in Vietnam that covers planning, budgeting, interministry and interlocal government cooperation, and delivery. Several studies on infrastructure development in Vietnam (Long et al. 2004; Nguyen and Dapice 2009) focus on one or two aspects of infrastructure development, rather than on an entire management system. The challenges of managing more extensive infrastructure networks require such an approach to focus the central government on planning, regulation, and policy elaboration and implementation; and to balance the roles of national, sectoral and local stakeholders.

1.6 Organization of the Book

Major contents of this report are organized in the following chapters:

Chapter 2 starts with a description of different goals of development. It is followed by an overview of some of the most prominent theories of economic development. These theories are categorized into three main clusters: early views, classical theories and contemporary theories. A number of economic development models are suggested by these theories. The most prominent models reviewed in this chapter are linear-stages-of-growth models, structural-change models, international-dependence models, neoclassical counterrevolution models, new growth theory and theory of coordination failure. Each model is reviewed with its tools and strategies as well as its major weaknesses. Among these models, implications of the theory of coordination failure for this study are highlighted and are further discussed in Chap. 5 of this report.

A review of the literature on construction and development is presented in Chap. 3. The relationship between infrastructure construction and economic development is reviewed through the contributions of construction to the economy as a whole, the contributions of construction to other sectors in the economy and the use of construction for macroeconomic stabilization. This chapter also discusses current concerns of the industry's performance, especially the infrastructure sector as an economic activity to stimulate economic growth in developing countries.

Chapter 4 provides an overview of different perspectives on the connection between international trade and economic development. This chapter focuses on the debate between neoclassical trade theory and dependency theory. Finally, this chapter highlights the importance of the integration of developing countries into the international trading system and related policies and challenges, including infrastructure development and renovation.

Chapter 5 presents a discussion on the intersection between trade, infrastructure construction and economic development. This chapter focuses on the theory of coordination failure and the role of government in infrastructure development for trade and economic growth. Consequently, this chapter discusses current issues in using public investments in infrastructure and the knowledge gap. This chapter also reviews the literature to identify major issues that could affect the efficiency of government-funded infrastructure in supporting trade and economic growth. Concerns discussed in this chapter are the basis for the formulation of the research hypotheses.

Chapter 6 provides an overview of economic development in Vietnam over the past decades. It is followed by a summary of the integration of Vietnam into the world economy. The contributions and development of the construction sector in Vietnam are also discussed. The next part of this chapter illustrates the inefficient use of public investments in infrastructure in planning and delivery in Vietnam. In addition, this chapter provides an understanding of the planning process at the national level, and major tasks of government agencies related to trade and infrastructure construction planning and development.

Chapter 7 discusses the research methodology adopted in this study. This chapter starts with a review of the use of qualitative and quantitative research strategies within the literature of public administration. This section is followed by a brief description of the research design selected for research in infrastructure planning. Details of approaches used for data analysis and data collection for this study are then discussed.

Chapter 8 focuses on data analysis and discussion. A model of the critical factors affecting the current use of public investments in infrastructure to support trade and economic growth in Vietnam is examined and validated. Results from both quantitative and qualitative data analyses are used to enrich the research findings.

Finally, Chap. 9 summarizes the subject matters of this study and major research findings. This chapter also presents research recommendations and validation. This chapter ends with research contributions, limitations and recommendations for future works.

References

Arnold, M. (2011). Infrastructure overview: The backbone of the global economy. Retrieved July 12, 2011, from http://campaign.etfsecurities.com/aem/clients/ETF001/images/Infrastructure March2011.pdf.

APEC Economic Committee (1997, June 18–20). *Infrastructure development in APEC.* Paper presented at the 1997 Public-Business/Private Sector Dialogue, Los Cabos, Mexico.

Batra, R. (1997). Marketing issues and challenges in transitional economies. *Journal of International Marketing, 5*(4), 95–114.

Burns, L. S., & Grebler, L. (1984). Is public construction countercyclical? *Land Economics, 60*(4), 367–377.

Davis, H. A. (Ed.). (2008). *Infrastructure finance: Trends and techniques.* London: Euromoney Books.

Demetriades, P. (2007). Transport infrastructure, trade and economic growth. In European Conference of Ministers of Transport (Ed.), *Report of the 134th Round Table on Transport Economics—Market Access, Trade in Transport Services and Trade Facilitation* (pp. 121–139). Paris, France: European Conference of Ministers of Transport.

Devarajan, S., Swaroop, V., & Zou, H. (1996). The composition of public expenditure and economic growth. *Journal of Monetary Economics, 37*(2), 313–344.

Estache, A. (2006). PPI partnerships vs. PPI divorces in LDCs. *Review of Industrial Organization, 29*(1), 3–26.

Fay, M., & Yepes, T. (2003). *Investing in infrastructure: What is needed from 2000 to 2010?* (World Bank Policy Research Working Paper No. 3102). Washington, D.C.: World Bank.

Flanagan, R., & Jewell, C. A. (2009). *Economics, urbanisation and infrastructure.* Geneva: FIDIC —International Federation of Consulting Engineers.

Flyvbjerg, B. (2007). Megaproject policy and planning: Problems causes, cures. *Environment and Planning B: Planning and Design, 34*, 578–597.

Flyvbjerg, B. (2008). Public planning of mega-projects: Overestimation of demand and underestimation of costs. In H. Priemus, B. Flyvbjerg & B. V. Wee (Eds.), *Decision-making on mega-projects: Cost-benefit analysis, planning and innovation* (pp. 120–144). Cheltenham, UK: Edward Elgar.

Gruneberg, S. L. (1997). *Construction economics: An introduction.* Basingstoke: Macmillan.

Harrod, R. F. (1948). *Towards a dynamic economics, some recent developments of economic theory and their application to policy*. London: Macmillan.

Hillebrandt, P. M. (2000). *Economic theory and the construction industry* (3rd ed.). Basingstoke: Macmillan.

Hoekman, B., Mattoo, A., & English, P. (Eds.). (2002). *Development, trade, and the WTO: A handbook*. Washington, D.C.: World Bank Publications.

Hossain, L. (2000). Infrastructure development planning in SEA: An analysis of the national telecommunications planning practice in Thailand. *Technovation, 20*(5), 275–285.

Ive, G., & Gruneberg, S. L. (2000). *The economics of the modern construction sector*. Basingstoke: Macmillan.

Kohsaka, A. (Ed.). (2007). *Infrastructure development in the Pacific region*. London: Taylor & Francis.

Long, N., Ogunlana, S., Quang, T., & Lam, K. (2004). Large construction projects in developing countries: A case study from Vietnam. *International Journal of Project Management, 22*(7), 553–561.

Marshall, T. (2011). Reforming the process for infrastructure planning in the UK/England 1990–2010. *Town Planning Review, 82*(4), 441–467.

Meier, G. M. (2000). The old generation of development economists and the new. In G. M. Meier & J. E. Stiglitz (Eds.), *Frontiers of development economics: The future in perspective* (pp. 13–50). Washington, D.C.: World Bank: Oxford University Press.

Mundial, B. (1994). *World development report 1994: Infrastructure for development*. Washington, D.C.: The World Bank.

Mustajab, M. (2009). Infrastructure investment in Indonesia: Process and impact (*dissertation*, University of Groningen, Nederlands.). Retrieved from http://irs.ub.rug.nl/ppn/318943859.

Nguyen, X. T., & Dapice, D. (2009). *Vietnam's infrastructure constraints* (Policy Dialogue Paper No. 3). Cambridge, MA: ASH Institute for Democratic Governance and Innovation, John F. Kennedy School of Government.

Nwankwo, S. (2000). Assessing the marketing environment in sub-Saharan Africa: Opportunities and threats analysis. *Marketing Intelligence and Planning, 18*(3), 144–153.

O'Sullivan, A., & Sheffrin, S. M. (2004). *Economics: Principles in action*. New Jersey: Prentice Hall.

Rostow, W. W. (1960). *The stages of economic growth: A non-communist Manifesto*. Cambridge: Cambridge University Press.

Schwab, K. (Ed.). (2008). *The global competitiveness report 2008–2009*. Geneva: World Economic Forum.

Schwab, K. (Ed.). (2009). *The global competitiveness report 2009–2010*. Geneva: World Economic Forum.

Schwab, K. (Ed.). (2010). *The global competitiveness report 2010–2011*. Geneva: World Economic Forum.

Sohail, M., & Cavill, S. (2008). Accountability to prevent corruption in construction projects. *Journal of Construction Engineering and Management, 134*(9), 729–738.

Tan, W. (2002). Construction and economic development in selected LDCs: Past, present and future. *Construction Management and Economics, 20*(7), 593–599.

Todaro, M., & Smith, S. (2009). *Economic development* (10th ed.). Boston: Addison Wesley.

Turin, D. A. (1969). *The construction industry: Its economic significance and its role in development*. London: University College, Environmental Research Group.

Venables, A. (2004). Infrastructure, trade costs and gains from international trade. In European Conference of Ministers of Transport (Ed.), *Report of the 130th Round Table on Transport Economics: Transport and International Trade* (pp. 57–82). Paris, France: European Conference of Ministers of Transport.

World Bank. (1984). *The construction industry, issues and strategies in developing countries*. Washington, D.C.: World Bank.

World Bank. (1994). *World development report 1994: Infrastructure for development*. Washington, D.C.: World Bank.

World Bank. (2010). Data by Country: Vietnam. Retrieved November 20, 2010, from http://data.worldbank.org/country/vietnam.

World Bank. (2011). Glossary. Retrieved August 7, 2011, from http://www.worldbank.org/depweb/english/beyond/global/glossary.html.

World Trade Organization. (2010). *Trade in natural resources* (World Trade Report).

Chapter 2
Theories of Economic Development

Abstract This chapter reviews some of the most prominent theories of economic development. These theories describe tools and strategies for making development goals achievable. The chapter starts with early views about the nature of economic prosperity. The chapter then reviews classical theories with four main clusters: linear stages of growth models; structural change models; international dependence models; and neoclassical counter-revolution models. Subsequently, contemporary theories of economic development, including new growth theory and theory of coordination failure, are reviewed. Finally, implications of the changes in the development thoughts and their importance in studying development problems in the developing countries conclude the chapter.

Keywords Economic development goals · Theories of economic development · Developing countries · Classical theories · Contemporary theories

2.1 Introduction

The problems of economic development, which are complex and multidimensional, have resulted in the development of a number of theories, explanations, arguments and assertions (World Bank 2000). The purpose of this chapter is to review some of the most prominent theories of economic development. These theories describe tools and strategies for making development goals achievable. The chapter starts with early views about the nature of economic prosperity. The chapter then reviews classical theories with four main clusters: linear stages of growth models; structural change models; international dependence models; and neoclassical counter-revolution models. Subsequently, contemporary theories of economic development, including new growth theory and theory of coordination failure, are reviewed. Finally, implications of the changes in the development thoughts and their importance in studying development problems in the developing countries conclude the chapter.

G. Dang and L. Sui Pheng, *Infrastructure Investments in Developing Economies*,
DOI 10.1007/978-981-287-248-7_2

2.2 Goals of Economic Development

2.2.1 *Growth of Gross National Product*

The goal of economic development in its simplest form is to create the wealth of a nation. Prior to the 1970s, rapid economic growth has been considered a good proxy for other attributes of development (Todaro and Smith 2009). Economic performance is measured by an annual increase in gross national product (GNP[1]) [an alternative measure is gross domestic product (GDP)]. For the purpose of comparability, GNP is expressed in a common currency, usually US dollars, and reported in per-capita terms to take into account the size of a nation's population (Jaffee 1998). The World Bank now replaces GNP per capita with gross national income (GNI) per capita to compare wealth among countries. The World Bank defines GNI as the sum of value added by all resident producers plus any product taxes (less subsidies) not included in the valuation of output plus net receipts of primary income (compensation of employees and property income) from abroad. Meanwhile, the World Bank still uses GDP in many other featured economic indicators (World Bank 2011).

However, the indicator is a measure of well-being and development exclusively based on material wealth. Improvements in welfare such as better health care, education and more housing for large parts of the poor population have not been captured. The experience of the 1950s and 1960s has shown that GNP growth would not necessarily result in a better life for a nation's population. The narrow goal of development (economic growth) induced nations to focus their energies narrowly on the rapid growth of national incomes (Todaro and Smith 2003). "To maximize income growth, environmental considerations were left to languish on the sidelines; the standard of living was often allowed to slide; large inequalities between classes, regions, and genders were ignored; and poverty was tolerated more than it should have been in the rush to generate maximum growth" (Basu 2000, p. 64). It was then scholars and policy-makers in most developing countries who realized that income growth was only one dimension of development; a new economic view of development has arrived.

2.2.2 *Quality of Life*

During the 1970s, the concern of millions of people living subsistence lives in poverty turned the attention of development economists to people's lives rather than their incomes. Many developing countries have experienced high growth rates of

[1] GNP is gross domestic product (GDP) plus incomes received by residents from abroad minus incomes claimed by non-residents. GDP is calculated as the value of the total final output of all goods and services produced in a single year within a country's boundaries (Soubbotina 2004).

per-capita income but little change in the living conditions of a large part of the population. By questioning whether it is the goal of development that per-capita income increases but poverty, inequality and unemployment are growing worse, Seers (1969) marked the change needed in setting development objectives. The goal of development during the period was thus not limited to economic growth but to concentrate on the reduction of poverty, inequality and unemployment (Seers 1979).

In the 1990s, economists increasingly recognized that it was the quality of life that determines whether people are from developing countries or not. Diseases, malnourishment and death that happen in the everyday lives of those from the developing countries changed the view of development goals dramatically. By then, like many scholars around the world, Stiglitz (1998) contributed to shift the development goals set by governments in developing countries to wider objectives, including improvements in income distribution, environment, health and education. A broader perspective of development goals is hence necessary as reflected in the World Bank's Development Report (1991, p. 4) as "to improve the quality of life. Especially, in the world's poor countries, a better quality of life generally calls for higher incomes—but it involves much more. It encompasses as ends in themselves better education, higher standards of health and nutrition, less poverty, a cleaner environment, more equality of opportunity, greater individual freedom, and a richer cultural life."

Sen's (1985, 1992, 1999) work perhaps has brought about the broadest perspective of development goals. According to Sen (1985), the ultimate goal of development is to enhance human capabilities, which is defined as "the freedom that a person has in terms of the choice of functionings, given his personal features (conversion of characteristics into functionings) and his command over commodities…" (Sen 1985, p. 13). Higher income is necessary but not sufficient in terms of quality of life. Under his approach, goals of economic development change from promotion of growth to promotion of well-being.

These changes in the definition of development goals posed the need to construct alternative composite indices to reflect quality of life. These indices should take into account not only money indicators but also non-monetary indicators to reflect the development levels achieved. There have been attempts to build indicators that measure the standard of living and quality of life, which focus on the quantitative and qualitative aspects: health, education, environment and material well-being (Berenger and Verdier-Chouchane 2007). Using Sen's (1985) approach, the Human Development Index (HDI) has been published annually since 1990 by the United Nations Development Programme as an attempt to provide an aggregate measure of life expectancy, education and income (Elkan 1995).

2.2.3 Sustainable Development

Increasingly, academics and societies realize the effects of human actions on the environment. On the way to achieve rapid economic growth, countries around the world have been exploiting their natural resource reserves at alarming rates.

Although early economists included the natural environment in their economic analysis, environmentalism only drew international attention in the 1960s (Pearce and Turner 1990). The relationship between development and environment has given birth to the sustainable development concept. The central idea of sustainable development is that global ecosystems and humanity itself can be threatened by neglecting the environment.

Environmental economists are concerned that the long-term neglect of the environmental assets is likely to jeopardize the durability of economic growth (Thampapillai 2002). Sustainable development therefore "involves maximizing the net benefits of economic development, subject to maintaining the services and quality of natural resources over time" (Pearce and Turner 1990, p. 24). Its concern is about balancing the objectives of economic growth and attending to environmental considerations.

In a broader sense, sustainable development is defined by the Brundtland Commission, formally the World Commission on Environment and Development, as "progress that meets the needs of the present without compromising the ability of future generations to meet their own needs" (World Commission on Environment and Development 1987, p. 8). Although this standard definition brings the term "sustainable development" into common use, it has created ambiguity in application (Redclift 1992; Daly 1996; Payne and Raiborn 2001). Much of the debate around the definition seeks to answer the two questions "What should be sustained" and "What should be developed" (Kates et al. 2008).

Today, sustainable development aims to improve the quality of life in a comprehensive manner, including economic prosperity, social equity and environmental protection. Economic, social, environmental and cultural aspects must be integrated in a harmonious manner to enhance the intergenerational well-being (World Bank 2003).

2.2.4 The Millennium Development Goals

Eight Millennium Development Goals (MDGs) were adopted by member countries of the United Nations in September 2000. The MDGs were developed to address the most pressing problems in developing countries, including poverty and hunger, primary universal education, gender equality, child health, maternal health, HIV/AIDS, environmental sustainability and global partnership. Member countries of the United Nations have committed themselves to end poverty and achieve other development goals by 2015. Quantitative targets of these goals were then assigned based on the past rates of international development achievements (United Nations 2011).

However, the MDGs were criticized for failing to include other critical objectives of development, such as improving legal and human rights of the poor, slowing global warming and leveraging the contributions of the private sector. Critics also argued that the MDG targets were not ambitious enough and were not prioritized (Todaro and Smith 2009).

The latest 2012 report showed a remarkable progress made by countries, including those in sub-Saharan Africa. The review maintained that the MDGs are still achievable. Increased supports from national governments, the international community, civil society and the private sector are considered necessary to meet the MDGs (United Nations 2012).

2.3 The Evolution of Economic Development Thoughts

2.3.1 Early Views About the Nature of Economic Society and Prosperity

Although development economics became established as a discipline within economics only in the 1950s, several early economists had written extensively about the nature of economic society and prosperity. Among them, Adam Smith and Karl Marx are the two most famous thinkers for their two opposite views on the nation's system of economic arrangements: one called capitalism and the other called socialism.

On the one hand, Adam Smith's (1976) (original work published in 1776) "The Wealth of Nations" focuses on the market. Adam Smith saw that division of labour could create more productive processes. The mechanism for enhancing the nation's wealth therefore is through specialization and exchange. Adam Smith argued that under competition, private investors while pursuing their own interests guided by the "invisible hand" would maximize national output and thus promote public interests. The "invisible hand" doctrine has become the foundation for the working of the market economy or capitalism (Skousen 2007). In the system, government interference is seen as inefficient in looking after economic activities. Meanwhile, free trade, private property and competition are seen as the foundations that would spur economic development, reduce poverty and bring on social and moral improvements of humankind. However, freewheeling capitalism is often criticized for bringing wealth only to the rich, whereas the poor get poorer.

On the other hand, Karl Marx in "Capital" (Marx 1933) (original work published in 1867) argued that the feasible system should be based on social or public ownership of property. Karl Marx emphasized that the wealth of the capitalists comes from the exploitation of the surplus value created by the workers. Hence, private property and free market were seen as causes of poverty for the many millions of workers. Therefore, private property should be completely abolished. A nation's economy should be planned and managed by the state to serve the interests of the masses. Marx believed that a revolution would be inevitable to break down the increasing concentration of the capitalists, and to establish socialism (Roemer 1988; Skousen 2007). But the socialism philosophy was not viable either. The historical experience of socialist economies showed little or even no improvement in the living conditions of the poor. The collapse of the Soviet Union in 1991 and the central planning paradigm appeared to demonstrate that the model would not provide the solution to poverty and inequality seen in human society (Meier 2000).

2.3.2 *Classical Theories of Economic Development*

2.3.2.1 The Linear Stages of Growth Models

The first generation of economic development models was formulated in the early years after the World War II. These early models focused on the utility of massive injections of capital to achieve rapid GDP growth rates. The two famous models are Rostow's stages growth model and the Harrod–Domar model (Todaro and Smith 2009).

Theorists of the 1950s and early 1960s viewed the process of development as a sequence of historical stages. This view was popularized by Rostow (Ingham 1995). Building on the historical pattern of the then developed countries, Rostow (1960) claimed that the transition from underdevelopment to development would pass through five stages: the traditional society, the preconditions for take-off, the take-off, the drive to maturity and the age of high mass consumption. The decisive stage is the take-off, through which developing countries are expected to transit from an underdeveloped to a developed state. Increasing rate of investments is considered to be necessary to induce per-capita growth. Like Rostow's stages growth model, the Harrod–Domar model emphasized that the prime mover of the economy is investments (Ghatak 2003). Every country therefore needs capital to generate investments. The principal strategies of development from the stage approach were commonly used by developing countries in the early post-war years. With a target growth rate, the required saving rate can then be known. If domestic savings were not sufficient, foreign savings would be mobilized.

Although Rostow (1960), Harrod (1948) and Domar (1947) were right about the important role of investments that is most closely correlated with the economic growth rate, this is not the only condition for a country to develop. The key weakness of these models lies in their simplifying assumptions. A single production function is simply assumed for all countries (Adelman 2000). Every economy is assumed to have the same necessary conditions and would pass through the same phasing, stage by stage. But that economic growth path, which historically had been followed by the more developed countries, is not the only one pathway. The development process is actually highly nonlinear (Chenery 1960; Chenery and Syrquin 1975). Countries may pursue distinct development paths (Morris and Adelman 1988). Economies may miss stages, or become locked in one particular stage, or even regress depending on many other complementary factors such as managerial capacities, and the availability of skilled labour for a wide range of development projects (Todaro and Smith 2009).

2.3.2.2 Structural Change Models

During most of the 1960s and early 1970s, economists generally described the development process as structural change by which the reallocation of labour from the agricultural sector to the industrial sector is considered the key source for economic growth. Two well-known representatives of this approach are the two-sector model (Lewis 1954), and the structural change and patterns of development (Chenery 1960).

In Lewis' (1954) two-sector model or theory of surplus labour, labour increasingly moves away from the agricultural sector to the industrial sector. However, with unlimited supply of labour from the traditional sector, these transferred workers continually received only subsistence wages. The excess of modern sector profits over wages and hence investments in the modern sector continued to expand and generate further economic growth on the assumption that all profits would be reinvested. Both labour transfer and modern sector employment growth were in turn brought about by output expansion in that sector. This process of modern sector self-sustaining growth and employment expansion facilitated the structural transformation from a traditional subsistence economy to a more modern developed economy to take place. Like the Harrod–Domar model, the Lewis model considered savings and investments to be the driving forces of economic development but in the context of the less developed countries. However, several Lewis' assumptions are not valid such as those relating to rural surplus labour, and the proportional rate of expansion in capital accumulation in the modern sector (Todaro and Smith 2009).

Although promoting the roles of savings and investments, the structural change and patterns of development analysis extended in comparison with the Lewis model. The analysis identified that the steady accumulation of physical and human capital is among conditions necessary for economic growth, apart from savings and investments. Moreover, the structural changes occurred not only in the two sectors but also in all economic functions, including the change in consumer demand from an emphasis on food and basic necessities to desires for diverse manufactured goods and services, international trade and resource use as well as changes in socioeconomic factors such as urbanization and the growth and distribution of a country's population. The most significant explanation of this approach was provided by Chenery (1960), Chenery and Taylor (1968), Kuznets (1971) and Chenery and Syrquin (1975).

By focusing on the pattern of development rather than theory, the structural change models may mislead policy-makers. Since the reallocation of labour from the agricultural sector to the industrial sector is considered the engine of economic growth, many developing countries implemented policies that often promote the industry and neglect agriculture. But the negative effects of policies that turned against that vital sector have come to be widely recognized (World Bank 2000). Criticisms of these models were reinforced by the fact that in many developing countries, poverty was prevalent. Following the pattern recommended by structural change economists, in the late 1960s, the attention of policy-makers began to shift towards an emphasis on human capital, i.e. education and health (Meier 2000). Then again, investments in health and education alone do not guarantee development. "In Sub-Saharan Africa, for example, life expectancy and school enrolment rates have increased dramatically in recent decades, but as a group the economies in the region have had slow and even negative growth since the early 1970s" (World Bank 2000, p. 16).

The structural change models focused on the pattern of development and hypothesized that the pattern was similar in all countries and was identifiable. However, empirical works, such as Chenery (1960), Chenery and Taylor (1968), and Chenery and Syrquin (1975), on the process of structural change does recognize that pattern of development can be different among countries, which is dependent on the countries' particular set of factors including "a country's resource endowment and size, its government's policies and objectives, the availability of external capital and technology, and the international trade environment" (Todaro and Smith 2009, p. 120).

2.3.2.3 International Dependence Models

The international dependence theory was very popular in the 1970s and early 1980s. The dependence theorists argued that underdevelopment exists because of the dominance of developed countries and multinational corporations over developing countries. The theory is considered an extension of Marxist theory (Hein 1992).

The poor countries are said to be dependent on the developed countries for market and capital. However, developing countries received a very small portion of the benefits that the dependent relationship brought about. The unequal exchange, in terms of trade against poor countries, made free trade a convenient vehicle of "exploitation" for the developed countries. Developed countries can exploit national resources of developing countries through getting cheap supply of food and raw materials. Meanwhile, poor countries are unable to control the distribution of the value added to the products traded between themselves and the developed countries (Cohen 1973; Dos Santos 1973). The growth of international capitalism and multinational corporations caused poor countries to be further exploited and more dependent on the developed countries. Poor countries therefore could not expect sustained growth from that dependence. Following the international dependence theory, developing countries should therefore end the dependence by breaking up their relationships with the developed world, as well as by closing their doors on the developed countries (Elkan 1995; Ghatak 2003; Ferraro 2008).

The models gained increasing support among the developing countries because of the limited results of the stages and structural change models. However, the failures of the model were clearly reflected in the developing countries that followed the autarky policy. These countries often experienced stagnant growth and finally decided to open their economies once again such as China, Tanzania and India (Ferraro 2008; Todaro and Smith 2009). Meanwhile, the experience of the newly industrialized economies of East Asia, namely Hong Kong, Singapore, Taiwan and South Korea, during the 1970s and 1980s showed that their success had been the result of emphasizing trade with the advanced industrial countries. The negative impacts of the policy of autarky rendered the theory out of favour in the 1980s (Hein 1992; Ferraro 2008).

2.3.2.4 Neoclassical Counter-Revolution Models

In the 1980s, neoclassical counter-revolution economists used three approaches, namely the free market approach, the new political economy approach and the market-friendly approach, to counter the international dependence model. In contrast with the international dependence model, these approaches mainly argued that underdevelopment is not the result of the predatory activities of the developed countries and the international agencies but was rather caused by the domestic issues arising from heavy state intervention such as poor resource allocation, government-induced price distortions and corruption (Meier 2000). As a response to public sector inefficiency, economists of the counter-revolution thinking, for example Bauer (1984), Lal (1983), Johnson (1971), and Little (1982), focused on promoting free markets, eliminating government-imposed distortions associated with protectionism, subsidies and public ownership.

Another strand of neoclassical free market thoughts called the traditional neoclassical growth theory actually originated from the Harrod–Domar and Solow models. Expanding the Harrod–Domar formulation, Solow neoclassical growth model stresses the importance of three factors of output growth: increases in labour quantity and quality (through population growth and education), increases in capital (through savings and investments) and improvements in technology (Solow 1956). Technological change in Solow's model is provided exogenously. Thus, with the same provided rate of technological progress, the growth rate would be expected to converge across countries. By opening up national markets, developing countries can draw additional domestic and foreign investments, thus increasing the rate of capital accumulation and returns on investments. Consequently, developing countries tend to converge to higher per-capita income levels (World Bank 2000).

Neoclassical economists focused on the market to find a way out for the developing countries. Policies of liberalization, stabilization and privatization therefore become the central elements of the national development agenda. Foreign trade, private international investments and foreign aid flowing into the developing countries are expected to accelerate economic efficiency and economic growth of these countries. Empirically, the models, however, did not bring about the expected results. The growth rates per capita have diverged among countries (Azariadis and Drazen 1990). Several African countries focusing on these issues achieved an average growth rate of only 0.5 % per year. With weak and inadequate legal and regulatory framework, not to mention the different institutional, cultural and historical context of the developing countries, free market in these countries fails to stimulate economic development (World Bank 2000).

2.3.3 Contemporary Theories of Economic Development

2.3.3.1 New Growth Theory

Endogenous growth or the new growth theory emerged in the 1990s to explain the poor performance of many less developed countries, which have implemented policies as prescribed in neoclassical theories. Unlike the Solow model that considers technological change as an exogenous factor, the new growth model notes that technological change has not been equal nor has it been exogenously transmitted in most developing countries (World Bank 2000).

New growth theorists (Romer 1986; Lucas 1988; Aghion and Howitt 1992) linked the technological change to the production of knowledge. The new growth theory emphasizes that economic growth results from increasing returns to the use of knowledge rather than labour and capital. The theory argues that the higher rate of returns as expected in the Solow model is greatly eroded by lower levels of complementary investments in human capital (education), infrastructure, or research and development (R&D). Meanwhile, knowledge is different from other economic goods because of its possibility to grow boundlessly. Knowledge or innovation can be reused at zero additional cost. Investments in knowledge creation therefore can bring about sustained growth. Moreover, the knowledge could create the spillover benefits to other firms once they obtained the knowledge. However, markets failed to produce enough knowledge because individuals cannot capture all of the gains associated with creating new knowledge by their own investments. Policy intervention is thus considered necessary to influence growth in the long term. The new growth models therefore promote the role of government and public policies in complementary investments in human capital formation and the encouragement of foreign private investments in knowledge-intensive industries such as computer software and telecommunications (Meier 2000).

Although the new growth theory helps to explain the divergence in growth rates across economies, it was criticized for overlooking the importance of social and institutional structures (Skott and Auerbach 1995). Its limited applicability lies in its assumptions. For example, it treats the economy as a single firm that does not permit the crucial growth-generating reallocation of labour and capital within the economy during the process of structural change. Moreover, there are many other factors which provide the incentives for economic growth that developing countries lack such as poor infrastructure, inadequate institutional structures and imperfect capital and goods markets (Cornwall and Cornwall 1994). Policy-makers will therefore need to pay careful attention to all of the factors that determine the changes and their impacts on the aggregate growth rate.

2.3.3.2 Theory of Coordination Failure

The foundation of the theory of coordination failure is the idea that the market may fail to achieve coordination among complementary activities. When complementaries exist, that is when returns of one investment depend on the presence or extent of other investments, there exist two scenarios. On the one hand, optimally, all investors as a whole are better off with all investments to be achieved at the same time. On the other hand, it would not make sense for an investor to take similar actions when he believes that others may not do the same as well. The market is said to have failed to coordinate investors' actions in this way. Coordination failure therefore leads the market to an (equilibrium) outcome inferior to a potential situation in which resources would be optimally allocated and all agents would be better off. As a result, underdevelopment equilibrium is possible (Hoff and Stiglitz 2000).

The theory of coordination failure became influential in the 1990s. However, it has a history of more than half a century. Coordination issues among complementary industries were first raised by Rosenstein-Rodan (1943). Like Rosenstein-Rodan (1943), early coordination failures economists Nurkse (1953) and Hirschman (1957) emphasized the role of the government to solve the problem. In order to reach an optimal level of coordination, the policy they recommended was a "big push"—a public-led massive investment program—which can cause complementarities to take place in the rest of the economy.

Like other early development models, "big push" strategies ran out of favour when the world witnessed the collapse of centrally planned economies and the slow growth, stagnation or worst results of state-led industrialization in the underdeveloped countries (Meier 2000). However, development economists have recently returned to emphasize the problem of complementarities between several conditions necessary for successful development to take place (Glăvan 2008). Hoff (2000), and Bowles Durlauf and Hoff (2006) described the economy as an ecosystem where the behaviour of one can affect the others'. The coordination failure among many different individuals lead the economy to multiple equilibria, but not all of them are good for every member of the economy, and some in fact are very undesirable. As a result, the market fails to coordinate everyone to achieve the optimal equilibrium. In other words, "A firm's productivity depends not only on its own efforts, and abilities, and on general economic conditions (for example, the macroeconomic environment and the legal system), but also on the actions of other firms, infrastructure, regulation and other public goods" (Rodriguez-Clare 2005, p. 3). In a similar vein, Rodrik (2004) also indicated that success or failure of an action could depend on its milieu.

In a market mechanism, there are uncertainties that a good equilibrium can be obtained. A bad equilibrium can exist when firms have pessimistic expectations and thus show their reluctance to invest, and consequently fail to coordinate their businesses. "And whereas in the past we thought the implication was that the economy would be slightly distorted, we now understand that the interaction of these slightly distorted behaviours may produce very large distortions. The consequence is that there may be multiple equilibria and that each may be inefficient" (Hoff and Stiglitz 2000, p. 390). The existence of coordination failure cannot therefore be disputed and has become important. When the market mechanism does not work, the active roles of the government need to be highlighted. According to coordination failure economists, in the multiple equilibria circumstances described above, the government can coordinate firms to move them into the domain of good equilibrium.

The theory of coordination failures offers some important overall lessons for policy-makers. The theory often highlights the problems of market failure that require selective government intervention to ensure that several things work well together at the same time. However, to get sustainable development underway is obviously not an easy task. The "big push" strategy is recommended recently by United Nations Development Programme (2005). The programme suggests that for developing countries to break out of the poverty trap, a big push of basic investments between now and 2015 in public administration, human capital and key infrastructure is necessary (United Nations Development Programme 2005).

However, the theory of coordination failure has been criticized for its overemphasis on the roles of government. Critics have asserted that the government is ineffective and could choose a bad policy (Killick 1976; Hoff and Stiglitz 2000). If a bad policy is implemented, it can push an economy into a bad equilibrium for years to come and even into a worse equilibrium than the one with which the country began (Hoff and Stiglitz 2000). Moreover, the policies recommended by coordination failure models lacked details of how the government can coordinate the economy. Policy-makers therefore need to be more cautious of these strategies to address coordination failure issues. The theory is further discussed in Chap. 5.

2.4 Summary

The review of the literature shows that there is increasingly a consensus that economic development is a multidimensional process that involves interactions among different goals of development and therefore would require systematically designed policies and strategies. Development issues are complex and multifaceted. There is no one single pathway for economic development that all countries can pursue. In the long term, the economic development process requires changes in policies to account for new emerging factors and trends. Designing these economic development policies also need to take into consideration the social, cultural, political systems and institutions as well as their changing interaction over time in a country.

Development strategies have changed remarkably over the past half century. Classical development economists often see underdevelopment as having a single cause. But history has demonstrated that focusing on one single factor alone cannot guarantee success in the development process. Capital formation (as emphasized in the linear stage growth models) is necessary but not sufficient. Structural change models that promoted industry but neglected agriculture also did not bring about the expected results. The international dependence models pursued an inward-looking model of development that promoted state-run production. On the contrary, the neoclassical free market counter-revolution is a different strand of thought that supported the role of the free market, privatization and export expansion. However, the contemporary models of development see the government and the market as complements, in which a certain extent of government intervention is required to ensure that desirable outcomes can be achieved in the presence of related market failures. These changes in development thoughts are shown in Fig. 2.1.

Although the ultimate goal of economic development goes beyond the growth of gross income (GDP, GNP or GNI) per capita, an understanding of the sources of growth is essential to achieve other objectives. Economic development is about growth plus organizational change (Hoff and Stiglitz 2000). Without growth, the change is unlikely to occur, since a country needs resources to realize other long-

Goals of development:

- Growth (of gross income per capita);
- Improvement in quality of life;
- Sustainable development;
- The Millennium Development Goals.

Development thoughts (prime movers of economic development):

- Early views:
  - Capitalism: free trade, private property and competition (Smith, 1776); and
  - Communism: social or public ownership of property, and independence of foreign capital and goods (Marx, 1933).
- Classical theories:
  - The linear-stages-of-growth models: savings and investments (Rostow, 1960; Harrod, 1948; Domar, 1947);
  - Structural-change models: transferring resources from low to high-productivity activities (from traditional/agricultural to modern /industrial sector) (Lewis, 1954; Chenery, 1960);
  - International-dependence models: withdrawal from the international economy and pursuit of self-sufficiency or autarky; and
  - Neoclassical counter-revolution models: liberalization, stabilization, and privatization.
- Contemporary theories:
  - New growth theory: knowledge or innovation;
  - Theory of coordination failure:
    - Underdevelopment as a coordination failure among complementary activities; and
    - Government intervention to move the economy to a preferred equilibrium.

Fig. 2.1 The evolution of development thoughts

term objectives. Growth and change will thus continue to be central to any development strategy.

The critical knowledge in finding the source of growth has been closely related to capital formation. However, as reviewed previously, the major weakness of the early theories is that they focused on finding the constraints in capital formation of one factor, such as physical capital or human capital, that limit economic growth. Hence, their solution is simply to increase investments in the factor identified. History has shown that the solution to a single cause does not always guarantee successful economic development. The solution is not simply an increase in that capital. More importantly, the focus should be on how to use the capital in an economy that consists of a combination of interrelated production processes. As recently realized by the contemporary development economists, especially by the theorists of coordination failures, the solution to obtain sustainable development underway is to make sure that several things work well simultaneously. Economic development is a complex process which involves causal relationships. One cannot risk overlooking these relationships as they lie at the centre of the development process. The theory of coordination failure thus served as a theoretical basis for connecting growth, trade and infrastructure construction later in this study. The theory itself is further discussed in Chap. 5.

References

Adelman, I. (2000). Fallacies in development theory and their implications for policy. In G. M. Meier & J. E. Stiglitz (Eds.), *Frontiers of development economics: The future in perspective* (pp. 103–134). Washington, D.C.: World Bank/Oxford University Press.

Aghion, P., & Howitt, P. (1992). A model of growth through creative destruction. *Econometrica, 60*(2), 323–351.

Azariadis, C., & Drazen, A. (1990). Threshold externalities in economic development. *The Quarterly Journal of Economics, 105*(2), 501.

Basu, K. (2000). On the goals of development. In G. M. Meier & J. E. Stiglitz (Eds.), *Frontiers of development economics: The future in perspective* (pp. 61–86). Washington, D.C.: World Bank and Oxford University Press.

Bauer, P. T. (1984). *Reality and rhetoric: Studies in the economics of development.* London: Weidenfield & Nicolson.

Berenger, V., & Verdier-Chouchane, A. (2007). Multidimensional measures of well-being: Standard of living and quality of life across countries. *World Development, 35*(7), 1259–1276.

Bowles, S., Durlauf, S. N., & Hoff, K. (Eds.). (2006). *Poverty traps.* Princeton, N.J.: Princeton University Press.

Chenery, H. B. (1960). Patterns of industrial growth. *The American Economic Review, 50*(4), 624–654.

Chenery, H. B., & Syrquin, M. (1975). *Patterns of development, 1950–1970.* London: Oxford University Press for the World Bank.

Chenery, H. B., & Taylor, L. (1968). Development patterns: Among countries and over time. *The Review of Economics and Statistics, 50*(4), 391–416.

Cohen, B. J. (1973). *The question of imperialism: The political economy of dominance and dependence.* New York: Basic Books.

Cornwall, J., & Cornwall, W. (1994). Growth theory and economic structure. *Economica, 61*(242), 237–251.

Daly, H. E. (1996). *Beyond growth: The economics of sustainable development.* Boston: Beacon Press.

Domar, E. D. (1947). Expansion and employment. *The American Economic Review, 37*(1), 34–55.

Dos Santos, T. (1973). The crisis of development theory and the problem of dependence in Latin America. In H. Bernstein (Ed.), *Underdevelopment and development: The third world today: Selected readings* (pp. 57–80). Harmondsworth: Penguin.

Elkan, W. (1995). *An introduction to development economics* (2nd ed.). New York: Prentice Hall.

Ferraro, V. (2008). Dependency theory: An introduction. In G. Secondi (Ed.), *The development economics reader* (pp. 58–64). London: Routledge.

Ghatak, S. (2003). *Introduction to development economics* (2nd ed.). London: Routledge.

Glăvan, B. (2008). Coordination economics, poverty traps, and the market process: A new case for industrial policy? *The Independent Review, 13*(2), 225–243.

Harrod, R. F. (1948). *Towards a dynamic economics, some recent developments of economic theory and their application to policy.* London: Macmillan.

Hein, S. (1992). Trade strategy and the dependency hypothesis: A comparison of policy, foreign investment, and economic growth in Latin America and East Asia. *Economic Development and Cultural Change, 40*(3), 495–521.

Hirschman, A. (1957). *The strategy of economic development.* New Haven: Yale University Press.

Hoff, K. (2000). Beyond Rosenstein-Rodan: the modern theory of coordination problems in development. In B. Pleskovic & N. Stern (Eds.), *Annual World bank conference on development economics* (pp. 145–188). Washington, D.C.: World Bank.

Hoff, K., & Stiglitz, J. E. (2000). Modern economic theory and development. In G. M. Meier & J. E. Stiglitz (Eds.), *Frontiers of development economics: The future in perspective* (pp. 389–485). Washington, D.C.: World Bank/Oxford University Press.

Ingham, B. (1995). *Economics and development.* New York: McGraw-Hill.

Jaffee, D. (1998). *Levels of socio-economic development theory* (2nd ed.). Westport, Connecticut: Praeger Publishers.
Johnson, H. (1971). A word to the third world: A Western economist's frank advice. *Encounter, 37*, 3–10.
Kates, R. W., Parris, T. M., & Leiserowitz, A. A. (2008). What is sustainable development? Goals, indicators, values, and practice. In G. Secondi (Ed.), *The development economics reader* (pp. 367–387). London: Routledge.
Killick, T. (1976). The possibilities of development planning. *Oxford Economic Papers, 28*(2), 161–184.
Kuznets, S. S. (1971). *Economic growth of nations: Total output and production structure*. Cambridge: Belknap Press of Harvard University Press.
Lal, D. (1983). *The poverty of "development economics"*. London: Institute of Economic Affairs.
Lewis, W. A. (1954). Economic development with unlimited supplies of labour. *The Manchester School, 22*(2), 139–191.
Little, I. M. D. (1982). *Economic development: Theory, policy, and international relations*. New York: Basic Books.
Lucas, R. E. (1988). On the mechanics of economic development. *Journal of Monetary Economics, 22*(1), 3–42.
Marx, K. (1933). *Capital*. London: J. M. Dent.
Meier, G. M. (2000). The old generation of development economists and the new. In G. M. Meier & J. E. Stiglitz (Eds.), *Frontiers of development economics: The future in perspective* (pp. 13–50). Washington, D.C.: World Bank/Oxford University Press.
Morris, C. T., & Adelman, I. (1988). *Comparative patterns of economic development, 1850–1914*. Baltimore: Johns Hopkins University Press.
Nurkse, R. (1953). *Problems of capital formation in underdeveloped countries*. Oxford: Basil Blackwell.
Payne, D. M., & Raiborn, C. A. (2001). Sustainable development: The ethics support the economics. *Journal of Business Ethics, 32*(2), 157–168.
Pearce, D. W., & Turner, R. K. (1990). *Economics of natural resources and the environment*. New York: Harvester Wheatsheaf.
Redclift, M. (1992). The meaning of sustainable development. *Geoforum, 23*(3), 395–403.
Rodriguez-Clare, A. (2005). *Microeconomic interventions after the Washington Consensus* (Working Paper No. 544). Washington, D.C.: Inter-American Development Bank.
Rodrik, D. (2004). *Industrial Policy for the twenty-first century* (CEPR Discussion Paper No. 4767). London: Centre for Economic Policy Research.
Roemer, J. E. (1988). *Free to lose*. Cambridge, M.A.: Harvard University Press.
Romer, P. M. (1986). Increasing returns and long-run growth. *Journal of Political Economy, 94* (5), 1002–1037.
Rosenstein-Rodan, P. N. (1943). Problems of industrialisation of Eastern and South-Eastern Europe. *The Economic Journal, 53*(210/211), 202–211.
Rostow, W. W. (1960). *The stages of economic growth: A non-communist manifesto*. Cambridge: Cambridge University Press.
Seers, D. (1969). The meaning of development. *International Development Review, 11*(4), 3–4.
Seers, D. (1979). The meaning of development, with a postscript. In D. Lehmann (Ed.), *Development theory: Four critical studies* (pp. 9–30). London: Cass.
Sen, A. (1985). *Commodities and capabilities*. Amsterdam: North Holland.
Sen, A. (1992). *Inequality re-examined*. Oxford: Clarendon.
Sen, A. (1999). *Development as freedom*. Oxford: Oxford University Press.
Skott, P., & Auerbach, P. (1995). Cumulative causation and the "new" theories of economic growth. *Journal of Post Keynesian Economics, 17*(3), 381–402.
Skousen, M. (2007). *The big three in economics: Adam Smith, Karl Marx and John Maynard Keynes*. New York: ME Sharpe Inc.
Smith, A. (1976). *An inquiry into the nature and causes of the wealth of nations*. Oxford: Clarendon Press.

Solow, R. M. (1956). A contribution to the theory of economic growth. *The Quarterly Journal of Economics, 70*(1), 65.

Soubbotina, T. P. (2004). *Beyond economic growth: An introduction to sustainable development* (2nd ed.). Washington, D.C.: World Bank.

Stiglitz, J. (1998). *More instruments and broader goals: Moving toward the post-Washington consensus*. Helsinki: Wider Annual Lecture United Nations University—World Institute for Development Economics Research.

Thampapillai, J. D. (2002). *Environmental economics: Concepts, methods and policies*. Australia: Oxford University Press.

Todaro, M., & Smith, S. (2003). *Economic development* (8th ed.). Boston: Addison Wesley.

Todaro, M., & Smith, S. (2009). *Economic development* (10th ed.). Boston: Addison Wesley.

United Nations. (2011). *Millennium goals*. Retrieved August 15th, 2011, from http://www.un.org/millenniumgoals/bkgd.shtml.

United Nations. (2012). *The millennium development goals report*. New York.

United Nations Development Programme. (2005). *UN millennium project—investing in development: A practical plan to achieve the millennium development goals*. London: Earthscan.

World Bank. (1991). *World development report 1991: the challenge of development*. Washington, D.C.: World Bank.

World Bank. (2000). *Entering the 21st century—World development report 1999/2000*. New York: Oxford University Press.

World Bank. (2003). *Sustainable development in a dynamic world—Transforming institutions, growth, and quality of life (World Development Report)*. Washington, D.C.: World Bank/Oxford University Press.

World Bank. (2011). Indicators. Retrieved August 15th, 2011 http://data.worldbank.org/indicator/NY.GNP.PCAP.PP.CD.

World Commission on Environment and Development. (1987). *Our common future*. New York: Oxford University Press.

Chapter 3
Construction and Economic Development

Abstract This chapter reviews the relationship between construction, especially infrastructure construction, and economic development. The relationship is discussed through the contributions of construction to the economy as a whole, the contributions of construction to other sectors in the economy and the use of construction for macroeconomic stabilization. Besides, some current concerns of the sector's performance as an economic activity to stimulate economic growth and how governments manage the development of the construction industry are also reviewed in the chapter.

Keywords Construction industry · Economic development · Infrastructure investments · Capital formation · Government institutions

3.1 Introduction

This chapter reviews the studies which evaluated the role of the construction sector, especially infrastructure, in the process of economic development. The chapter first describes some basic concepts and characteristics of the construction industry. The remaining parts of the chapter will demonstrate the critical contributions of construction to the economy. Contributions of the construction industry come from the linkages between the construction sector and the economy as a whole, and the intersectoral linkages between construction and other sectors as well. The use of infrastructure investments as a tool for the government to stabilize the economy also shows the sector's key position in the national development strategy. Despite these prominent roles, some current concerns of the sector's performance as an economic activity to stimulate economic growth will also be reviewed in the chapter.

G. Dang and L. Sui Pheng, *Infrastructure Investments in Developing Economies*,
DOI 10.1007/978-981-287-248-7_3

3.2 Some Basic Concepts of the Construction Industry

3.2.1 Construction as an Economic Activity

Generally, there are three levels to define construction within the literature. At one extreme, construction is referred to as an economic activity that involves the entire construction process from producing raw and manufactured building materials and components, and providing professional services such as design and project management, to executing the physical work on site. In this view, construction is an economic activity that crosses over all three economic sectors: primary sector that involves the extraction of natural resources; secondary sector that involves the manufacture of building materials and components, and the transformation of these materials into finished buildings; and tertiary sector that involves the provision of consultancy services such as project management, design and structural engineering (Gruneberg 1997). From this angle of approach, the construction process actually starts long before the physical work on site that transforms materials and design into the complete buildings, structures and facilities.

At the other extreme, construction is conceived as an economic activity that focuses only on the last stage of the construction process which is the physical work carried out on the production site. From this perspective, all services such as project management, planning and design as well as the offsite manufacture and supply of building materials are excluded. One typical example of this kind of definition is the one provided in the International Standard Industrial Classification of all Economic Activities (Department of Economic and Social Affairs 2008). According to this classification, project management for construction, architectural and engineering activities, and the manufacture of building materials are listed under different categories other than construction. This manner of classification is considered convenient for statistical purposes (Ofori 1990). Following this classification, construction is perceived as an economic activity directed to the new work, renovation, repair or extension of buildings, structures and other heavy constructions such as roads, bridges, dams and so forth (Department of Economic and Social Affairs 2008). Accordingly, only the work force working on the construction site is defined as the labour force of the construction industry (Ive and Gruneberg 2000).

There is one more way of defining the construction industry, which is somewhat in between the two above-mentioned extremes. In this context, construction is referred to as the production process of the built environment including various activities from conception through design to execution. The objects of the built environment include buildings and other fixed structures (Ive and Gruneberg 2000). In other words, the construction industry is "a sector of the economy which, through planning, design, construction, maintenance and repair, and operation, transforms various resources into constructed facilities" (Moavenzadeh 1978, p. 98). As such, the construction industry encompasses all firms or organizations, who professionally engage in the construction process, from those providing consultancy services in the planning, design, supervisory and managerial services to those carrying out execution work on site such as the general contractors and builders (Colean and Newcomb 1952; Ofori 1990). These aspects are directly related to the capacity of infrastructure delivery. This definition is thus more relevant to this study.

3.2.2 Construction Products

This section looks at the classification of the construction industry's products and their characteristics. The classification of products is necessary as each type of output represents different kinds of supply and demand, which will be affected by different factors in the market (Ofori 1990). According to the International Standard Industrial Classification of all Economic Activities (Department of Economic and Social Affairs 2008), complete products of the construction industry consist of dwellings, office buildings, stores and other public and utility buildings, farm buildings and so on, or heavy constructions such as highways, streets, bridges, tunnels, railways, airfields, harbours and other water projects, irrigation systems, sewerage systems, industrial facilities, pipelines and electric lines, sports facilities and so on. More generally, final outputs of the construction industry are grouped into housing, infrastructure, industrial and commercial buildings, and repair and maintenance (Wells 1984; Gruneberg 1997). Another way to classify construction outputs is to group them into civil engineering work, buildings, and repairs and maintenance. Civil engineering work includes transport facilities, telecommunication and power networks, water supplies and so on. Buildings include housing and the remainder (hospitals, schools, offices, factories, hotels and agricultural buildings) (Wells 1985).

Construction's final outputs share certain common characteristics with each other, such as their custom-built nature, immobility, complexity, durability and costliness (Colean and Newcomb 1952; Moavenzadeh 1978). These products are also often distinguished from others in terms of time lag, labour-intensive operations, site production and temporary organizations (Ofori 1990; Koskela 2003). It is not difficult to envisage products of the construction industry in these terms. Construction products are not mass produced commonly. Each construction product is only constructed after the client decides to procure it. Custom-built nature or the one-of-a-kind nature of construction outputs is featured by different clients with different needs and priorities, by different sites and by different views of designers (Warszawski 1990). Products of the construction industry are immobile; they are fixed to the site where they will be used. The products are constructed on site as well. As a result, the construction process depends not only on the natural conditions of the site but also on the local resources such as the local labour force, local subcontractors, local building materials suppliers and local regulations. Complexity of construction products follows from their custom-built nature and site production. The high diversity of customers' requirements and site conditions results in a wide range of materials, technological requirements, design solutions and construction methods of varying degrees of complexity. Another reason for the complexity is the number of different teams from different parties involved in the construction works (Ofori 1990). Each team with different professional practices, specializing in different operations, often leads to a complicated organization in the construction process. The organization is only temporary for a particular product.

Construction products are durable. They are built to generate a flow of services for a very long time. Civil engineering works and buildings thus often last for generations. Except for catastrophes caused by human or nature, a construction product is only demolished when it becomes obsolete and no one wants to use it, which means it is no longer economical to maintain (Colean and Newcomb 1952; Gruneberg 1997). Durability and complexity are the main reasons why the construction of civil engineering works and buildings is capital intensive, involves a large number of labour and takes a long time to complete. Delays are also often unavoidable. Delays can be caused by unexpected natural events, technical, financial, procedural and contractual problems (Ofori 1990).

3.2.3 Structure of the Construction Industry

By nature of the work, the structure of the construction industry can be characterized by type of product, size of contract, complexity and geographical location of finished products (World Bank 1984). By type of product, the construction industry consists of construction firms specializing in residential, industrial and commercial buildings, and infrastructure or civil engineering works. The construction industry can also be structured by size of contract and degree of complexity. Large construction firms usually undertake large contracts with high degree of complexity. The technical as well as managerial skills are the two major advantages that large construction firms have over the small firms. This explains that the potential competitors for large civil engineering works are often large construction firms. Geographical location is also an important determinant of the structure of the construction industry. Construction firms are usually market located (Gruneberg 1997). Because construction product is characterized by site production, most construction firms need to consider the transportation costs of materials, equipment and labour to the production site. Beyond a certain geographical area, these costs become excessive in relation to other costs, thus rendering the project's profitability unattractive. Only for large contracts would the proportion of transportation costs in total costs become small enough, so that it is still profitable for firms to undertake the jobs farther afield. For very large projects, firms might set up a local office to reduce transportation costs, but then, these firms will need to consider the attendant overheads associated (Hillebrandt 2000).

By pattern of production organization, this study focuses on the formal sector rather than the informal one. Within the formal sector, the structure of the construction industry can be classified into public and private firms, and domestic and foreign firms. Public firms are state-owned enterprises, which are often protected by the government. The establishment of these public enterprises is essential to take on works which are unattractive to private firms or as required by the government, or perform as a source of providing the necessary trained labour for the development of the construction industry. Meanwhile, private contractors contribute to the development of the construction industry by their competitiveness, efficiency and flexibility. One more noticeable point in the structure of the construction industry is the participation of international contractors. Their presence can make the competition more difficult for domestic firms. However, domestic firms could become stronger in the process. The structure of the construction industry can be quite different among countries depending on the social and economic environments such as the development state of the economy, government policies and the traditions of doing business (World Bank 1984).

3.3 Major Economic Factors Influencing Construction Activity

3.3.1 Instability of Demand

Construction demand is inherently volatile. Wide fluctuations in construction outputs are common. Studying a number of countries with different income from various parts of the world, the World Bank (1984) indicated that the fluctuation of construction outputs was more than half of the manufacturing sector and the economy as a whole. A study of the European region by the United Nations (1976) came to the same conclusion that the construction industry experienced stronger movements of business cycle than other economic sectors. These fluctuations are partly due to the nature of construction demand and partly due to the nature of construction products. Demand for construction outputs is considered derived demand from other economic activities (United Nations 1976; Hillebrandt 2000). Outputs of the construction industry are investment goods, which are produced to facilitate the creation of other consumption goods and services. In other words, the construction industry itself cannot create the demand for its outputs (Nam and Tatum 1988). Demand for construction is therefore largely dependent on the business activities of other goods and services that the construction products help create. Following the ups and downs of the economy, the demand for construction can go through the movement of a business cycle earlier and more strongly than other sectors because of the nature of its products (Riggleman 1933; United Nations 1976). A construction product is immobile and is often constructed only after the client realizes the demand. This nature of construction products makes the industry unable to stock up products for sale. Hence, when there is a sudden increase in demand, the industry cannot respond to the demand quickly (United Nations 1976; Nam and Tatum 1988). The fluctuations of construction demand can be very large, thus affecting the economy considerably. The immobility and the durability properties also make construction demand geographically dependent. Demand for construction within a geographical area will at some time in the future experiences a substantial fall and suffers saturation when much of the demand in the area is met and the existing facilities are still economical to maintain (Colean and Newcomb 1952; Hillebrandt 2000).

Apart from the above-mentioned factors, the stages of economic development as well as the structural changes also have significant influences on the growth rate of demand for construction as well as the compositions of demand for construction over time. A study by the United Nations (1976) pointed out that the fluctuations in construction activity, with reference to those of other economic sectors, tend to be greater in developing than in developed countries. In developing countries, the demand for infrastructural facilities is of the greatest proportion compared with demand for other types of facilities. Since in the initial stages of development, infrastructure is of great importance in creating the framework for the economy, spending on building new infrastructure in these stages is relatively heavy. As a result, the share of construction expenditure on public works and public utilities such as harbours, railways, roads and electricity in the less developed countries is often higher than that in the more developed countries (Lewis 1955). Structural changes such as changes in development goals, priorities of economic growth among sectors and changes in demographics can also impact the demand for construction (United Nations 1976). The World Bank (World Bank 1994) observed that as countries develop, infrastructure must adapt to support changing patterns of demand.

3.3.2 Main Construction Inputs

Main inputs in construction include labour, equipment and materials. The characteristics of each component are discussed further below.

3.3.2.1 Labour

Since construction products are frequently labour intensive, labour is one of the major cost components in construction (Ive and Gruneberg 2000). The construction labour force can be divided into the major categories of administrative, managerial, professional and technical staff, and workers. Demand for labour is directly affected by the characteristics of demand for construction. In line with the cyclical fluctuations in construction outputs, the level of employment in the industry can vary noticeably (United Nations 1976). As a result, a large proportion of construction labour force is paid on a project-by-project basis, rather than on a permanent basis, except for a small number of administrative and managerial staff. Furthermore, outsourcing labour through subcontractors also makes the employment in construction increasingly temporary and insecure (International Labour Office 2001). At the bottom of the supply chain, construction workers have low job security and need for mobility (Moavenzadeh 1978). Careers in construction are also less attractive because of the high rate of accidents (Jensen 1983). The industry therefore has to compete with other sectors of the economy for manpower by increasing wage level (Ofori 1990). Labour can also be imported if the wages of the imported labour plus transport costs can be less than the local wages (Gruneberg 1997). In addition, the price of construction labour is affected by minimum wage laws and union agreements, which would raise the costs of hiring unskilled labour substantially (World Bank 1984).

The labour market of the construction industry is characterized by the shortage of skilled labour, especially in the developing countries. In the developed countries, the skilled workers account for about 50 % of the total labour force (Moavenzadeh 1978). However, the construction industry in these countries has apparently been experiencing a shrinking workforce, thus a reduction in the number of the skilled labour (Ofori 1990). Meanwhile, in the developing countries, the shortage of skilled labour is more serious. The labour markets in these countries are largely unstructured and relatively unorganized, which rely heavily on a relatively untrained workforce (Gruneberg 1997). In public works construction, the shortage of skilled labour occurs at both levels, the supervisory personnel and skilled workers (Moavenzadeh 1978). The inability of the industry to attract workers and to invest in training them has seriously affected the productivity and quality of construction products and hence the ability of contractors to satisfy the clients' needs (International Labour Office 2001). One of the solutions to the problem is to improve vocational training in the domestic construction industry. While this solution has proved to be successful in the developed countries, the non-availability or inadequacy of educational and training facilities is still a problem in the developing countries (Rao 1983; Edmonds and Miles 1984). Another way to reduce the dependence on labour is through the use of labour-saving construction technology (i.e. through the use of equipment-intensive technology). However, this solution has proved to be inappropriate in the developing countries as will be illustrated in the next section.

3.3.2.2 Equipment and Construction Technology

The intensity of equipment used in construction projects depends on construction technology. Construction technology reflects the level of mechanization used in the method of construction. The use of construction technology can vary across projects. This is because there is a certain degree of technological flexibility for the same construction work. For the same construction work, designers and contractors can choose between the two construction methods: one that is highly mechanized and the other that is fully manual (Moavenzadeh 1978). The choice of the contractor depends on the constraints of time and budget. If time is of the essence, then the highly mechanized method may be used. However, if the total cost of using plant and equipment plus wages for technical personnel and semi-skilled workers is much higher than the total labour cost of minimally trained unskilled workers used in the fully manual method, the latter may be used when the budget is limited (Ofori 1990). In developed countries where labour has become more costly compared to other inputs, the use of equipment-intensive technology has been more common than in developing countries where the cost of plant and equipment is high and labour is still cheap and abundant (International Labour Office 2001). In other words, the choice of technology is predominately affected by the prevailing prices of labour and equipment (World Bank 1984).

In a research study by the World Bank in 1971 (as cited by World Bank 1984), the role of labour-intensive technology in the development of infrastructure in developing countries has proved to be important. The results of the study showed that labour-intensive technology in capital-scarce, labour-abundant economies were technically and economically feasible for government-run labour-intensive civil construction programmes. With improvements at the managerial and technical level, labour-intensive methods could be fully competitive with equipment-intensive methods in terms of quality and productivity (World Bank 1984). Since labour continues to be abundant in developing countries, more serious efforts from the government will be essential to make the labour-intensive option more attractive for a wide range of construction activities.

3.3.2.3 Building Materials

Building materials make up a major component of total construction costs (Ive and Gruneberg 2000). The price level of building materials thus directly affects the construction output value. The price of a particular type of materials depends on the market forces, demand and supply of that building material. Demand for a particular building material in turn depends on client's tastes and preferences, and level of income; local building standards and codes; and the choices made by designers and contractors. The choices made by designers and contractors vary according to the experience of the designers, the familiarity of contractors with the technology involved in the use of the materials and the availability of human resource with the required skills (Ofori 1990).

The supply of a particular building material is affected by the availability of raw materials; the technology for extracting and processing raw materials; the environmental impacts related to the production process; and the government policy. The availability of raw materials directly determines the supply level of building materials. Some countries that lack raw materials have to import them. Apart from the lack of raw materials, building materials supply may be unable to meet local demand because of the low production capacity and low quality, especially in the developing countries. Production capacity and quality of building materials in turn depend on the technology of extracting and processing the raw materials used (Ofori 1990). The technology used in the developing countries is generally older than in the developed countries. In line with older technology, environmental impacts in the developing countries are also more serious than that in the developed countries. A growing awareness of these environmental impacts has made governments in the developing countries restrict the production of low-grade materials that uses the old technology and support the import of more modern technology from the developed countries (International Cement Review Research 2007, 2008).

Apart from market forces, the price of building materials may also be controlled by the government. To facilitate the local construction industry, the government can fix the price levels and provide subsidies for some materials, especially for the materials that are used in public sector projects. The government can encourage or discourage the import of some building materials by changing import duties, and taxes, or quotas on them (World Bank 1984; International Cement Review Research 2007, 2008).

3.3.3 Sources of Finance

Financing production in construction is quite different from that in manufacturing. The production in construction is financed largely by clients, rather than the contractors, the producers of construction products. At the top of the payment chain in the construction industry are the contractors, and at the bottom are the subcontractors and the suppliers. Contractors receive advance payments and periodic payments to mobilize resources (labour, equipment and materials) necessary for the construction projects. Cash flow of subcontractors and suppliers is therefore also affected by the payment procedure. The source of client's finance is often in the form of bank loans backed by the constructed facility itself. Although the major source of finance comes from clients, the contractors in some countries can obtain bank loans from special banks, which are established to assist the financing of construction works and investments in the construction industry. Other forms of financial assistance that help contractors get started in heavy construction works can include supplier credit, direct loans from commercial banks and the establishment of companies that lease and hire equipment (World Bank 1984).

Contractors' financial concerns can be different across the construction phases and the technology employed. The initial stages of mobilization require a large amount of cash, especially for contractors using the equipment-intensive construction technology to purchase the equipment. Meanwhile, cash flow at the later stages of an uneventful construction operation is generally more stable. Delayed payments and uncertainty of cash flow are common in the construction industry (World Bank 1984). The problem has continually raised concerns in construction industries throughout the world, although the problem can be more serious in the developing countries. Payment problems were considered a factor that can hamper the healthy development of the construction industry (Wu et al. 2008).

3.4 Construction and Aggregate Output

3.4.1 Capital Formation

The productive capacity of an economy is usually described in terms of the complete utilization of factors of production, that is, the full employment of the factors of production (labour and capital). Hence, the productive capacity determines the aggregate supply or national output of an economy in the long run. If there are changes in the total capital stock, the fixed amount of national output will change accordingly. One way for an economy to expand production or grow is to invest in capital stocks (human and physical resources). The relationship has long been recognized in economic theory. According to the Harrod-Domar model briefly reviewed in Chap. 2, net investment (I), which is defined as the change in the capital stock (ΔK), is necessary for the economy to grow. Besides investments in new factories, machinery, equipment, and materials, investments in economic infrastructure—roads, electricity, communication and the like—also increase the physical capital stock of a nation, thus expanding national output level. Investments in economic infrastructure are considered a supplement to other physical capital. For example, a farmer may increase the total output of the crops by an investment in a new tractor, but without adequate transport facilities, this extra product cannot be available in local commercial markets, and thus, his investment may not add anything to national food production. As a result, assuming there were only two products, products A and B, produced by the economy at a given technology level, increases in infrastructure resources would make it possible to shift the production possibility curve outward uniformly from P_1P_1 to P_2P_2 as shown in Fig. 3.1 (Todaro and Smith 2009).

Infrastructure can also raise productivity of other factors of production. For example, an installation of a new irrigation system can improve the quality of farmland, thus raising productivity per hectare and shifting the production possibility curve outward (Todaro and Smith 2009). The higher productivity in turn

Fig. 3.1 Effect of increases in infrastructure resources on the production possibility frontier

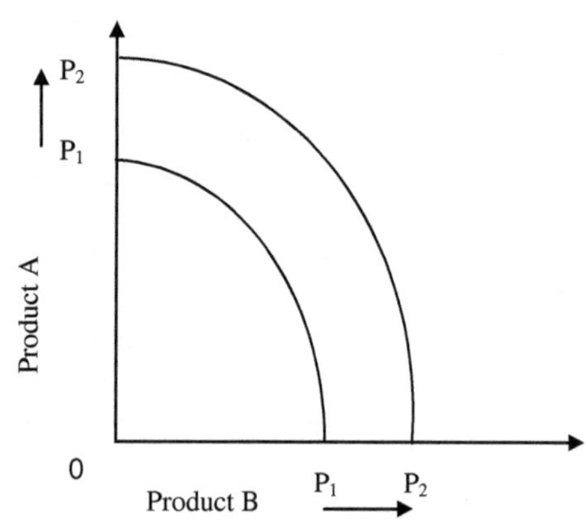

attracts more resources (private investments) into production, which further contribute to higher levels of output, profitability of production, income and employment in these sectors. The process of investing in the physical capital stocks of an economy, including infrastructure, is measured by the gross fixed capital formation (GFCF).

There has been a huge amount of literature analyzing the positive correlation between aggregate output growth and investments in infrastructure. In the end 1980s and early 1990s, many empirical studies on the returns on total infrastructure investments indicated high rates of returns in developed countries such as the USA (Aschauer 1989; Munnell 1990), Sweden (Berndt and Hansson 1991) and Canada (Wylie 1996). There have also been studies focusing on a specific type of infrastructure. In Baum and Korte (2002), transport was demonstrated to foster economic development by enhancing the efficiency of the economic system in both the goods and employment markets. Easterly and Rebelo (1993) studied historical time series and cross-country data of 28 developed countries. The study showed a highly and consistently positive correlation between transport and communication investments, and economic growth rate. Studies aimed at the developing countries have also been conducted by Canning and Fay (1993), and Canning (1998). Throughout these studies, transportation and telephone systems appear to promote economic growth. For example, Canning and Fay (1993) examined the rates of returns of the transportation (road and rail) infrastructure and telephone systems from 1960 to 1980 in 104 countries at different income levels. The cross-sectional analysis showed that both transportation and telephone systems had large effects on growth rates, with high rates of returns; especially for countries with lower ratios of infrastructure to output, the rates of returns were higher than 40 %. Many studies addressed the impacts of inadequate infrastructure on economic development as well. For example, in a study of economic development in Africa by Mountjoy and Hilling (1988), low rate of capital formation was argued to have a clear link with poverty in this continent. The study by the United Nations (1990) indicated that infrastructural development was an essential part of economic development in Asia and the Pacific region. As a result, this has been interpreted as a causal linkage, suggesting that infrastructure investment does appear to lead economic growth by improving the capacity as well as the efficiency of the economy.

As an industry supplying the physical infrastructure facilities, construction thus has a profound impact on economic development. GFCF by construction is the total value of all additions within a year, including new construction and all capital alterations or extensions, which significantly improve upon the utility or extend the life of the building or works to fixed capital. Repair and maintenance are not included in the indicator (Turin 1969; Ofori 1990). Many studies have emphasized that about half of the investments in GFCF in most developing countries comes from construction (Lewis 1955; Turin 1978; World Bank 1984; Wells 1985; Ofori 1988; Gruneberg 1997; Hillebrandt 2000; Ruddock and Lopes 2006). It is thus understandable that the construction industry plays a dominant role in a country's rate of GFCF (Lewis 1955; Gruneberg 1997). The cross-sectional analysis of 87 countries at different per capita GDP levels during the period of 1955–1965 by

Turin (1969) showed a strong linear correlation between the logarithms of per capita formation by construction and per capita GDP. The share of capital formation in construction in GDP also increased with per capita GDP (Turin 1969). Further studies by Strassmann (1970), Turin (1973) and Wells (1986) all argued that during the periods of accelerating economic growth, the construction industry needs to grow faster than the economy as a whole. A study by the World Bank (1994) also confirmed that construction has to expand fast enough to create adequate infrastructure and productive facilities for economic growth. Inadequate construction capacity is therefore a constraint on capital investment programmes (Wells 1986).

3.4.2 National Income

Construction like any other economic activities can contribute to national income by creating income. Based on the definition by the United Nations, Ofori (1990) explained that value added by construction is the gross output value at producer's prices less the value of all industry's current purchases from other enterprises. It is important to note that according to the narrowly defined scope of the construction industry, all industry's current purchases from other enterprises are excluded, which are the value of input materials and components, costs of hiring plant, costs of goods sold in the same conditions as purchased, legal and other professional fees, and payments made for repair and maintenance undertaken by others on the construction firm's own assets. Value added therefore is actually the sum of salaries and wages of employees, interest on borrowed capital, net rent, profit and allowance for depreciation (Ofori 1990). According to data provided by the World Bank (1984), the value added-to-output ratios of the construction industry in most developing countries were higher than those of manufacturing over the period of 1970–1980. In some countries, the ratios of the construction industry could be as high as 60 % (World Bank 1984).

The linkage between the value added by construction as a share in GDP and per capita GDP has long been recognized. According to Turin (1969) and Strassmann (1970), there is a strong linear correlation between the logarithms of per capita value added by construction and per capita GDP. The share of value added by construction as a percentage in GDP also increases as per capita GDP increases. The share of the valued added by construction as a percentage to GDP was found to be around 3–5 % for developing countries and 5–8 % for more developed countries over the period of 1955–1965 (Turin 1969). The results were later confirmed by many other studies such as those completed by Edmonds and Miles (1984), Wells (1985, 1986), Ofori (1988), Low and Leong (1992), Chen (1998) and more recently Lopes (2012). Although these studies focused more on the static view of the close relationship between construction activity and economic growth, these studies do note that construction's role in the economy would decline when the economy reaches the middle income stage (Strassmann 1970) or when the volume of construction products was sufficient to raise the productive capacity of the economy at a steady growth rate (Wells 1986). The contribution of the construction industry required for a steady economic growth as suggested by Edmonds (1979) should be 5 % of GDP. Lopes et al. (2002) and Lopes (2012) also demonstrated that when the share of construction value added in GDP was around 4–5 %, the economy would enter a period of sustained growth. The level should be interpreted within the long-term trend rather than the annual fluctuation (Lopes et al. 2002).

The dynamic views of construction in the national economy have been further examined. These views argued that as a country develops, the share of construction in total GNP as well as the total construction output actually follows an inverted U shape (e.g. Bon 1992, 2000). That is, in the initial development stages, the share of construction output in GNP increases, but starts to decline in the more mature stages. Thus, in the long run, the volume of construction output will decline accordingly. Therefore, the contribution of the construction industry to economic growth is not indefinite (nonlinear relationship). Bon (1992, 2000) observed that there is a point at which the basic major infrastructure is put in place and the need for new construction gradually becomes less. However, Ruddock and Lopes (2006), Lopes (2009) and Goh (2009) argued that in the more mature stages of development, construction output declines only in relative terms, not in absolute terms. That is, in the more developed countries, construction still grows but at slower rates than the economy. The inverted U-shaped relationship between the share of construction in GNP and GNP per capita proposed by Bon (1992, 2000) (as shown in Fig. 3.2) was further confirmed by Crosthwaite (2000) and Yiu et al. (2004). These observations implied that capacity expansion in construction is more important for the developing countries than for the developed countries. The feature of the activity in the industry is also different between the developing and developed countries. Repair and maintenance capacity become important in the developed countries, whereas the industry's capacity for new build as well as the development of capacity for repair and maintenance is a major focus of the developing countries (Bon and Crosthwaite 2000).

As a result, policy implications within the literature have focused on the measures to expand the capacity of the construction industry in the developing countries. The development strategies for the construction industry are consequently

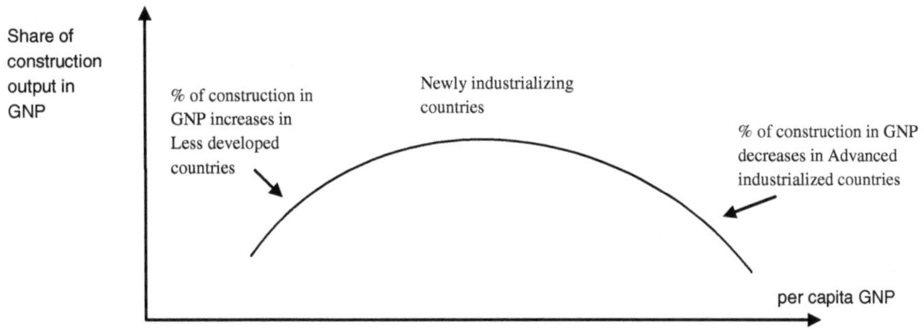

Fig. 3.2 Share of construction in GNP and per capita GNP

centred on the schemes to remove the constraints of the industry's production factors such as labour, materials, capital and technology. By building up an adequate domestic capacity, the construction industry is expected to drive economic growth. The policy recommendations also include creating an institutional environment that facilitates a competitive business environment for an efficient construction industry. At the macroeconomic level, monetary and interest rate policies are expected to deal with the distortions in factor prices, the limits to the availability of foreign exchange for the purchase of imported inputs and related issues such as foreign exchange rates and import tariffs. Meanwhile, fiscal policies related to government spending and taxes on public construction works are expected to encourage employment and to stabilize the construction activity. Throughout these studies, the role of governments has been demonstrated by using these policies to directly or indirectly influence construction activity to stimulate economic growth.

3.4.3 Other Observations

Further empirical studies have shown that the contribution of construction in the economy as suggested by Turin (1969) as well as Bon (1992, 2000) is not consistent across countries, especially in the developing countries. The data provided by the World Bank (1984) indicated that construction value added in most developing countries over the period of 1970–1980 varied between 3 and 8 % of GDP. Studying the data of 75 countries in 2003, Ruddock and Lopes (2006) found that the share of gross value added by construction varied considerably from over 2 % to over 10 % across countries in the same category of GDP per capita. Meanwhile, a study of the share of construction in GFCF by Lewis (2009) found that the figure was much lower than 50 % as suggested by previous studies. The average proportion of construction in GFCF from 1970 to 2006 for developed countries was around 26 % and for the developing countries was around 23 %. There was also a trend of divergence among the developing countries. The divergence trend was explained by the fact that most infrastructure in the developing countries was financed by the government and thus was directly affected by different government-expenditure policies (Lewis 2009).

The problem of the direction of the causality between construction and GDP has also been analysed. However, the results of these empirical studies were inconclusive. On the one hand, the study using data from Hong Kong by Tse and Ganesan (1997) indicated that the causality ran from GDP to construction activity. On the other hand, the study using data from Western Europe by Wilhelmsson and Wigren (2009) showed that the causality effect was weak in the case of infrastructural construction in the long run. Meanwhile, Chan (2001), in studying the linkage between construction and other economic sectors in Singapore, found a bidirectional causal relationship between construction activities and GDP. In the case of China, the study by Zheng and Liu (2004) also indicated a bidirectional casual relationship: construction investment had a strong short-run effect on economic growth, whereas economic growth had a long-term effect on construction. Lewis (2009) indicated that the relationship between construction and the national economy of Trinidad and Tobago changed over time under different circumstances: during the economic upturn, the economy led construction; and during the economic downturn, construction drove the economy.

In striking contrast to a number of the above studies, there have been studies concerning the negative impacts of the construction industry in the economy. The concern was first raised by Drewer (1980). Analyzing the data of the United Nations Economic Commission for Europe (ECE) region between 1970 and 1976, Drewer (1980) observed that more construction does not necessarily result in higher economic growth when resources are misallocated. Drewer (1997) found that the relationship between construction and the economy is unstable and the uncontrolled expansion of the construction industry could negatively impact the economy. The pressures generated by the expansion of the construction industry may push up the costs of inputs (such as labour, materials), affect the availability of financial capital for other uses and intensify environmental stress. As a result, the overexpansion of construction activity may affect macroeconomic stability by generating inflationary pressures and misallocating as well as wasting resources. The negative impacts of overexpansion of construction activity may considerably offset the real growth of the economy (Drewer 1997; Ng et al. 2008). Similarly, studying construction industry in China regional economy, Han and Ofori (2001) found different results of development programmes and that appropriate policies for the development of the construction industry would be key to stimulate economic growth.

3.5 Construction and Sectoral Output

The ability of the construction industry to stimulate economic growth also comes from the strong linkages between construction and other sectors in the economy. The construction industry is one of the top four out of twenty economic sectors in terms of intersectoral, backward and forward linkages (Riedel and Schultz 1980). The important role of construction through a complex set of interrelationships was also highlighted by Ofori (1990).

3.5.1 Backward Linkages

Since large quantities of building materials and components are purchased from a large number of supply industries, an expansion of the construction industry can stimulate the expansion of these industries through backward linkages. The impact can be significantly large because much of the building materials can be provided by relatively unsophisticated labour-intensive domestic resources and by basic industries such as cement and steel manufacturing (World Bank 1984). Many input–output analyses (Park 1989; Pietroforte and Bon 1995; Bon and Yashiro 1996; Bon et al. 1999) have demonstrated the strong backward linkages. Consequently, should the value added by construction take into consideration the backward linkages such as the manufacture of building materials and components, the value added can account for a considerable proportion in GDP. The value added through backward linkages could be up to 55 % of the value of construction purchases of materials and services from other industries (Kirmani 1988). However, it is important to note that the value added will be high when locally produced inputs are used. Therefore, while imported materials may be necessary in the short run, an economy needs to develop its local supply industries in the long run if it aims to increase value added. These issues have been addressed in many studies (Drewer 1980; Wells 1986).

3.5.2 Forward Linkages

The indices of forward linkages of the construction industry are relatively less extensive than that of the industry's backward linkages (Bon and Minami 1986; Bon and Pietroforte 1990; Pietroforte and Bon 1995; Bon and Yashiro 1996; Bon et al. 1999). Nevertheless, the magnitude of the forward linkages can be more significant since the demand for construction outputs is derived demand from all other sectors of the economy. As a result, how these sectors react to the change in the level of construction activity, and vice versa, ultimately affect the economy.

The construction industry through its outputs, especially infrastructure, can have impacts on outputs of other sectors. Many studies have shown the strong linkages between infrastructure and sectoral outputs. The strong linkages stem from the fact that all other sectors of the economy use the flow of services generated by infrastructure, which in turn can lead to growth in the production of other economic sectors in two ways: through the reduction in costs of intermediate inputs from infrastructure services such as transport, water and electricity; and through the increase in the productivity of other factors of production. As a result, the supply of infrastructure services can raise the profitability of production, level of returns, output, income and employment of other sectors. In studies by Binswanger et al. (1987) and Binswanger et al. (1993), roads and electricity supply were found to have a strong positive effect on aggregate agricultural output, as well as the growth of farm investments. Antle (1983) found that spending on transport and communication services was a significant determinant of differences in aggregate agricultural productivity across countries. The study by Chhibber (1988) concluded that public goods and services, including infrastructure, were the determinant of agricultural outputs in the developing countries; meanwhile, in more the developed countries which had better basic infrastructure, the determinant was price. Elhance and Lakshmanan (1988), Kranton (1991) and Lee and Anas (1992) examined the contribution of infrastructure to growth through reductions in costs and found that infrastructure could be a major share of business expenses. The unreliable delivery of infrastructure services and lack of access to infrastructure services could result in multiple economic costs such as direct costs of production delays, loss of perishable raw materials or outputs, costs of under-utilization of productive capacity and so on. All these important contributions of infrastructure on other sectors of the economy highlight the crucial role of construction's supply capacity in raising other sectoral outputs. However, little is known about the reverse flow of influence, which is how the adaptive capacities of other economic sectors can affect the construction sector. The influence is equally important since overexpansion of the construction sector and misallocation of resources in the economy might stem from the lack of understanding of this effect.

3.6 Construction and Macroeconomic Stabilization

The contribution of construction in the economy has also been measured by construction employment as a share of total employment (Strassmann 1970; Turin 1978; Wells 1985). Turin (1978) suggested that since employment in construction correlated positively with economic growth, there is a potential use of construction to generate sustained employment. Turin (1978) also argued that the misuse of construction by the government as a cheap way to absorb unskilled unemployment through ill-planned public work programmes would only damage the healthy development of the construction industry.

Numerous data have shown that there have been dramatic increases in construction employment in most developing countries (International Labour Office 2001). The fact that the construction industry is more labour intensive than many other industries, especially relative to manufacturing (Hillebrandt 2000), makes the industry a traditional focus of employment generation policy in many countries through labour-intensive public works projects. The labour created in these projects in turn spends the income from the employment on other locally produced goods and services, thereby stimulating demand in the rest of the economy through the multiplier effect. The employment multiplier effect also makes public works a more promising instrument for moderating the business cycle over others such as credit or tax incentives offered to the private sector. As a result, during periods of slack demand and high unemployment rate, infrastructural construction projects funded by the government are often implemented as a counter-cyclical instrument (Gruneberg 1997; Hillebrandt 2000). Similarly, the government can stabilize the economy by postponing these projects during boom periods (Burns and Grebler 1984). Through fiscal policy such as changing the amount of public expenditure or interest rates of loans financing these infrastructure projects, the government can generate desired changes in the economy, which has been the subject of many studies (Lea 1973; United Nations 1976; Lange and Mills 1979; Ofori 1988; Hillebrandt 2000; Howes and Robinson 2005). The recent 2008–2009 economic crisis has witnessed many countries relying considerably on construction spending to jump-start the economy and spur employment. Many countries around the world have included construction of infrastructure such as railways, highways, airports and power grids into their fiscal stimulus packages (Donnges 2009).

However, there are several difficulties in realizing the expected results. One of the difficulties is the time lag between government action and the actual construction works from the project start to completion. As a result, these projects can be counterproductive if the government cannot foresee problems far in advance (Hillebrandt 2000). Other difficulties might come from technical considerations when the government wants to postpone infrastructural construction. Technical problems make it difficult and costly to cancel infrastructural construction in midstream (Burns and Grebler 1984). An empirical study of the regional data of the USA by Duffy-Deno and Eberts (1991) indicated that these difficulties caused infrastructure to only have short-run multiplier effects.

In many poor countries, the objective of public programmes is usually a combination of poverty alleviation, employment generation and the provision of infrastructure. These public programmes have been carried out for decades in South Asia and Africa (World Bank 1994). However, the results of these programmes were varied. On the one hand, as reviewed by Kessides (1992), the emphasis of traditional public work programmes on short-term benefits (through quick creation of employment and assets) has dampened the long-term economic benefits, which could be derived from more careful project selection, creation of higher quality assets, and more emphasis on the training of workers. On the other hand, as reviewed by McCutcheon (2001), large-scale programmes for employment-intensive construction and maintenance have successfully created employment without compromising cost, quality or time. The experience suggests that these programmes should be integrated into other long-term development programmes to generate the greatest benefits.

3.7 Government Institutions for Construction Industry Development

The role of the construction industry in the economy recognized by academics has led to the recommendation of creating a government agency which specifically deals with the development of the construction industry. Having observed the critical role of construction in the developing economies, Turin (1978) suggested the establishment of such a government department. The roles and functions of such an agency were furthered analysed by Ofori (1985). Several countries have realized the constraints of the domestic construction industry and have established sophisticated agencies to guide the industry towards development. The Construction Industry Development Board of South Africa was established in 2000 after the 1997 Green Paper identified some of the challenges facing the industry (Construction Industry Development Board of South Africa 2000). In Australia, after a report prepared by the Joint Working Party (consisting of the major contractors, the

Australian Federation of Construction Contractors, the National Building and Construction Council and the National Public Works Committee) to provide a report on the difficulties facing the industry in the late 1980s, the government outlined the Construction Industry Reform Strategy, which led to the establishment of the Construction Industry Development Agency and Construction Industry Development Council in 1992 (Commonwealth of Australia Law 1992; Royal Commission into the Building and Construction Industry 2002). The Construction Industry Development Board of Singapore was formed in 1984 to deal with the then pressing issues of the industry (Ofori 1993). Some other examples of countries that have established the bureau or boards to respond to problems in the construction industry include Indonesia, India and Malaysia. The government agencies are often under ministerial control. Members of these agencies are usually appointed by the Ministry and compose of government officials, academics and private consultants to monitor and manage the sector. Generally, the creation of these agencies is to address the issues concerning the weaknesses of the local construction industry. The common functions of the agencies relate to formulating, monitoring, managing policies, standards, programmes and initiatives to deal with the shortage of local skilled labour, materials, low-output productivity and quality, or the low competitiveness of the local construction industry. Other focus might be the issues relating to sustainability, regulatory reforms, new procurement procedures and measurement of the industry's performance (Ofori 1993).

Other government agencies, professional bodies and trade associations, and academic institutions can also impact construction industry development (Miles and Neale 1991; Ofori 1994; Tan 2002; Milford 2012). Construction activities can be influenced by other government bodies such as the Ministry of Finance (through public spending and fiscal policies and so on), the Ministry of Trade and Industry (through raw materials, manufacturing regulations, prices, tariff, taxes and so on), and the Ministry of Labour (employment regulations, site safety, wage policies and so on). The roles of professional bodies and trade associations were also emphasized (Miles and Neale 1991; Ofori 1994). These institutions can enhance the competiveness of the construction industry by providing their members necessary skills through course accreditation, and supporting professional development through data services, computerized libraries and similar (Ofori 1994). In this context, academic institutions implement research programmes, providing the industry with pertinent views about shortcomings in construction, and new techniques or approaches to overcome these shortcomings.

By establishing these institutions to attend to pressing issues in the construction industry, governments have demonstrated an active commitment to construction industry development. The agency that is specifically concerned with construction industry development can be considered the core agency for this purpose. To function effectively, the core agency needs to coordinate with other government bodies and organizations in connection with the performance of its tasks. However, the intriguing question is how the coordination of the duties of these institutions towards the changing problems facing the industry can be effectively managed. The question cannot be answered without more research on the relationship between construction and other economic activities.

3.8 Conclusions

Much of the existing empirical literature on the linkage between construction and economic growth seeks to capture this effect through observation of the relationship between increases in the construction outputs (measured by the share of construction in GFCF or value added) and some measures of aggregate output (GDP or GNP). Many of the findings from these studies have demonstrated the positive and significant relationship between the construction industry and economic growth in the developing countries. However, the conclusions have been questioned when further studies on this relationship were conducted for more developed countries. The results of these studies indicate that the relationship between economic development and the construction industry appears to be more complicated. Moreover, whether growth in the share of the value of construction activities in total GDP is a cause or a consequence of economic growth is not clear. However, there exists a general consensus on the underlying pattern of the relationship between GDP per capita and the share of construction in total GDP. This means that in the initial development stages of an economy, the share of the construction industry in GDP increases at a faster rate, then levels off and finally declines at higher levels of economic development. This suggests that in developing countries, construction is still a crucial factor for consideration when economic policies in these countries are being formulated.

Since construction involves high stakes in the developing economies, most of the policy implications recommended within the literature focus on stimulating construction outputs and expanding the domestic construction capacity. However, the preceding discussion also suggests that there are high potential costs from overexpansion in the sector. Consequently, construction might contribute to the economy in the short run, but offset the real growth of the economy in the long run. Since demand for construction is derived demand from other sectors of the economy, construction can only contribute to the economy when there is sufficient complement and basic productive level of other sectors to absorb the construction outputs. Further expansion of the construction outputs, especially infrastructure, beyond the adaptive capacity of the economy will only waste national resources. The knowledge suggests that focus on additional investment and project financing for infrastructure development may not bring about the desired impacts on economic development if these projects are not linked to broader economic development. Appropriate policies for the development of construction outputs should thus be those coordinated with other sectors in the economy (Dang and Low 2011).

References

Antle, J. M. (1983). Infrastructure and aggregate agricultural productivity: International evidence. *Economic Development and Cultural Change, 31*(3), 609–619.

Aschauer, D. A. (1989). Is public expenditure productive? *Journal of Monetary Economics, 23*(2), 177–200.

Baum, H., & Korte, J. (2002). Report by H. Baum and J. Korte (Germany). In European Conference of Ministers of Transport (Ed.), *Report of the 119th Round Table on Transport Economics—Transport and Economic Development* (pp. 5–49). Paris, France: European Conference of Ministers of Transport.

Berndt, E. R., & Hansson, B. (1991). *Measuring the contribution of public infrastructure capital in Sweden* (NBER Working Paper No. 3842). Cambridge, MA., USA: National Bureau of Economic Research.

Binswanger, H., Yang, M.-C., Bowers, A., & Mundlak, Y. (1987). On the determinants of cross-country aggregate agricultural supply. *Journal of Econometrics, 36*(1–2), 111–131.

Binswanger, H. P., Khandker, S. R., & Rosenzweig, M. R. (1993). How infrastructure and financial institutions affect agricultural output and investment in India. *Journal of Development Economics, 41*(2), 337–366.

Bon, R. (1992). The future of international construction: Secular patterns of growth and decline. *Habitat International, 16*(3), 119–128.

Bon, R. (2000). *Economic structure and maturity*. Aldershot: Ashgate.

Bon, R., & Crosthwaite, D. (2000). *The future of international construction*. London: Thomas Telford.

Bon, R., & Minami, K. (1986). The role of construction in the national economy: A comparison of the fundamental structure of the US and Japanese input-output tables since World War II. *Habitat International, 10*(4), 93–99.

Bon, R., & Pietroforte, R. (1990). Historical comparison of construction sectors in the United States, Japan, Italy and Finland using input-output tables. *Construction Management and Economics, 8*(3), 233–247.

Bon, R., & Yashiro, T. (1996). Some new evidence of old trends: Japanese construction, 1960–1990. *Construction Management and Economics, 14*(4), 319–323.

Bon, R., Birgonul, T., & Ozdogan, I. (1999). An input-output analysis of the Turkish construction sector, 1973–1990: A note. *Construction Management and Economics, 17*(5), 543–551.

Burns, L. S., & Grebler, L. (1984). Is Public Construction Countercyclical? *Land Economics, 60* (4), 367–377.

Canning, D. (1998). A database of world stocks of infrastructure, 1950–1995. *World Bank Economic Review, 12*(3), 529–547.

Canning, D., & Fay, M. (1993). *The effect of transportation networks on economic growth*. New York: Columbia University, Department of Economics.

Chan, S. L. (2001). Empirical tests to discern linkages between construction and other economic sectors in Singapore. *Construction Management and Economics, 19*(4), 355–363.

Chen, J. J. (1998). The characteristics and current status of China's construction industry. *Construction Management and Economics, 16*(6), 711–719.

Chhibber, A. (1988). Raising agricultural output: Price and nonprice factors. *Finance and Development, 25*(2), 44–48.

Colean, M. L., & Newcomb, R. (1952). *Stabilizing construction: The record and potential* (1st ed.). New York: McGraw-Hill.

Commonwealth of Australia Law. (1992). Construction Industry Reform and Development Act 1992. Retrieved January 4, 2010, from http://www.comlaw.gov.au/ComLaw/Legislation/Act1.nsf/asmade%5Cbytitle/A31918FB03D36FAFCA256F720017C47F?OpenDocument.

Construction Industry Development Board of South Africa. (2000). Background and history. Retrieved January 4, 2010, from http://www.cidb.org.za/about/background/default.aspx.

Crosthwaite, D. (2000). The global construction market: A cross-sectional analysis. *Construction Management and Economics, 18*(5), 619–627.
Dang, T. H. G., & Low, S. P. (2011). Role of construction in economic development: Review of key concepts in the past 40 years. *Habitat International, 35*(1), 118–125.
Department of Economic and Social Affairs. (2008). *International standard industrial classification of all economic activities*. New York: United Nations.
Donnges, C. (2009). *Addressing unemployment and poverty through infrastructure development as a crisis response strategy*. Paper presented at the Impact of the global economic slowdown on poverty and sustainable development in asia and the pacific. Retrieved from http://www.adb.org/documents/events/2009/Poverty-Social-Development/infrastructure-and-rescue-packages-Donnges-paper.pdf.
Drewer, S. (1980). Construction and development: A new perspective. *Habitat International, 5* (3–4), 395–428.
Drewer, S. (1997). Construction and development: Further reflections on the work of Duccio Turin. In School of Building and Real Estate—NUS (Ed.), *Proceedings of the First International Conference on Construction Industry Development: Building the Future Together* (Vol. 1, pp. 132–139). Singapore: National University of Singapore.
Duffy-Deno, K. T., & Eberts, R. W. (1991). Public infrastructure and regional economic development: A simultaneous equation approach. *Journal of Urban Economics, 30*, 329–343.
Easterly, W., & Rebelo, S. (1993). Fiscal policy and economic growth. *Journal of Monetary Economics, 32*(3), 417–458.
Edmonds, G. A. (1979). The construction industry in developing countries. *International Labour Review, 118*(3), 355–369.
Edmonds, G. A., & Miles, D. W. J. (1984). *Foundations for change: Aspects of the construction industry in developing countries*. London: Intermediate Technology.
Elhance, A. P., & Lakshmanan, T. R. (1988). Infrastructure-production system dynamics in national and regional systems: An econometric study of the Indian economy. *Regional Science and Urban Economics, 18*(4), 511–531.
Goh, B. H. (2009). Construction and economic development of four most competitive economies in the world: A comparison. *International Journal of Construction Education and Research, 5* (4), 261–275.
Gruneberg, S. L. (1997). *Construction economics: An introduction*. Basingstoke: Macmillan.
Han, S. S., & Ofori, G. (2001). Construction industry in China's regional economy, 1990–1998. *Construction Management and Economics, 19*(2), 189–205.
Hillebrandt, P. M. (2000). *Economic theory and the construction industry* (3rd ed.). Basingstoke: Macmillan.
Howes, R., & Robinson, H. (2005). *Infrastructure for the built environment: Global procurement strategies*. Oxford: Butterworth-Heinemann.
International Cement Review Research. (2007). Concentration is the key. *International Cement Review,* Aug 2007, 30–45.
International Cement Review Research. (Aug, 2008). Strength through adjustment. *International Cement Review,* pp. 32–40.
International Labour Office. (2001). *The construction industry in the twenty-first century: Its image, employment prospects and skill requirement*. Geneva: International Labour Organization.
Ive, G., & Gruneberg, S. L. (2000). *The economics of the modern construction sector*. Basingstoke: Macmillan.
Jensen, R. (1983). Workers' compensation claims relating to heat and cold exposure. *Professional Safety, 28*(9), 19–24.
Kessides, C. (1992). *Note on public works employment schemes: A review of studies*. Washington, D.C.: Infrastructure and Urban Development Department, World Bank.

Kirmani, S. S. (1988). *The construction industry in development: Issues and options.* Washington, D.C.: Infrastructure and Urban Development Department, World Bank.

Koskela, L. (2003). Is structural change the primary solution to the problems of construction? *Building Research and Information, 31*(2), 85–96.

Kranton, R. E. (1991). *Transport and the mobility needs of the urban poor: An exploratory study* (World Bank Discussion Paper INU No. 86). Washington, D.C.: Infrastructure and Urban Development Department, World Bank.

Lange, J. E., & Mills, D. Q. (1979). *The construction industry: Balance wheel of the economy.* Lexington, MA: Lexington Books.

Lea, E. (May, 1973). Stop-go and the building industry. *Building,* pp. 115–118.

Lee, K. S., & Anas, A. (1992). *Impacts of infrastructure deficiencies on Nigerian manufacturing: Private alternatives and policy options* (Infrastructure and Urban Development Department Report No. 98). Washington, D.C.: Infrastructure and Urban Development Department, World Bank.

Lewis, W. A. (1955). *The theory of economic growth.* London: Allen & Unwin.

Lewis, T. M. (2009). Quantifying the GDP–construction relationship. In L. Ruddock (Ed.), *Economics for the modern built environment* (pp. 34–59). London: Taylor & Francis.

Lopes, J. (2009). Investment in construction and economic growth: A long-term perspective. In L. Ruddock (Ed.), *Economics for the modern built environment* (pp. 94–112). London: Taylor & Francis.

Lopes, J. (2012). Construction in the economy and its role in socio-economic development: Role of construction in socio-economic development. In G. Ofori (Ed.), *New perspectives on construction in developing countries* (pp. 40–71). UK: Spon Press.

Lopes, J., Ruddock, L., & Ribeiro, F. L. (2002). Investment in construction and economic growth in developing countries. *Building Research and Information, 30*(3), 152–159.

Low, S. P., & Leong, C. H. Y. (1992). A revisit to turin's paradigm. construction and development in the 1970s and 1980s. *Habitat International, 16*(3), 103–117.

McCutcheon, R. (2001). An introduction to employment creation in development and lessons learned from employment creation in construction. *Urban Forum, 12*(3–4), 263–278.

Miles, D., & Neale, R. (1991). *Building for tomorrow: International experience in construction industry development.* Geneva: International Labor Organization.

Milford, R. (2012). Institution building in construction industries in developing countries. In G. Ofori (Ed.), *New perspectives on construction in developing countries* (pp. 109–132). UK: Spon Press.

Moavenzadeh, F. (1978). Construction industry in developing countries. *World Development, 6*(1), 97–116.

Mountjoy, A. B., & Hilling, D. (1988). *Africa, geography and development.* London: Hutchinson.

Munnell, A. H. (Jan/Feb, 1990). Why has productivity growth declined? Productivity and public investment. *New England Economic Review,* pp. 3–22.

Nam, C., & Tatum, C. (1988). Major characteristics of constructed products and resulting limitations of construction technology. *Construction Management and Economics, 6*(2), 133–147.

Ng, S. T., Chan, A. P. C., Chiang, Y. H., Kumaraswamy, M. M., Lam, P. T. I., Lee, P. K. K., et al. (2008). *Reinventing the Hong Kong construction industry for its sustainable development (final report).* Hong Kong: Construction Industry Institute.

Ofori, G. (1985). Managing construction industry development. *Construction Management and Economics, 3*(1), 33–42.

Ofori, G. (1988). Construction industry and economic growth in Singapore. *Construction Management and Economics, 6*(1), 57–70.

Ofori, G. (1990). *The construction industry: Aspects of its Economics and Management.* Singapore: NUS Press.

Ofori, G. (1993). *Managing construction industry development: Lessons from Singapore's experience*. Singapore: Singapore University Press.

Ofori, G. (1994). Practice of construction industry development at the crossroads. *Habitat International, 18*(2), 41–56.

Park, S. H. (1989). Linkages between industry and services and their implications for urban employment generation in developing countries. *Journal of Development Economics, 30*(2), 359–379.

Pietroforte, R., & Bon, R. (1995). An input–output analysis of the Italian construction sector, 1959–1988. *Construction Management and Economics, 13*(3), 253–262.

Rao, A. N. (Ed.). (1983). *The Construction industry in Asia: A survey with a study*. Japan: Asian Productivity Organization.

Riedel, J., & Schultz, S. (1980). Construction and building materials industry in developing countries. *Economics, 21*, 36–64.

Riggleman, J. R. (1933). Building cycles in the United States, 1875–1932. *Journal of the American Statistical Association, 28*(182), 174–183.

Royal Commission into the Building and Construction Industry. (2002). *Recent reviews of the building and construction industry* (Discussion Paper No. 9). Australia: Royal Commission into the Building and Construction Industry.

Ruddock, L., & Lopes, J. (2006). The construction sector and economic development: The 'Bon curve'. *Construction Management and Economics, 24*(7), 717–723.

Strassmann, W. P. (1970). The construction sector in economic development. *Scottish Journal of Political Economy, 17*(3), 391–409.

Tan, W. (2002). Construction and economic development in selected LDCs: Past, present and future. *Construction Management and Economics, 20*(7), 593–599.

Todaro, M., & Smith, S. (2009). *Economic development* (10th ed.). Boston: Addison Wesley.

Tse, R. Y. C., & Ganesan, S. (1997). Causal relationship between construction flows and GDP: Evidence from Hong Kong. *Construction Management and Economics, 15*(4), 371–376.

Turin, D. A. (1969). *The construction industry: Its economic significance and its role in development*. London: University College, Environmental Research Group.

Turin, D. A. (1973). *Construction and development* (2nd ed.). London: University College, Environmental Research Group.

Turin, D. A. (1978). Construction and development. *Habitat International, 3*(1–2), 33–45.

United Nations. (1976). *Long-term prospects and policies in the construction sector*. New York: United Nations.

United Nations. (1990). *Economic and social survey of Asia and the Pacific 1990*. New York: United Nations.

Warszawski, A. (1990). *Industrialization and robotics in building: A managerial approach*. New York: Harper & Row.

Wells, J. (1984). The construction industry in the context of development: A new perspective. *Habitat International, 8*(3–4), 9–28.

Wells, J. (1985). The role of construction in economic growth and development. *Habitat International, 9*(1), 55–70.

Wells, J. (1986). *The construction industry in developing countries: Alternative strategies for development*. London: Croom Helm.

Wilhelmsson, M., & Wigren, R. (July 09, 2009). The robustness of the causal and economic relationship between construction flows and economic growth: Evidence from Western Europe. *Applied Economics,* (iFirst), pp. 1–10.

World Bank. (1984). *The construction industry, issues and strategies in developing countries*. Washington, D.C.: World Bank.

World Bank. (1994). *World development report 1994: Infrastructure for development*. Washington, D.C.: World Bank.

Wu, J., Kumaraswamy, M., & Soo, G. (2008). Payment problems and regulatory responses in the construction industry: Mainland china perspective. *Journal of Professional Issues in Engineering Education and Practice, 134*(4), 399–407.

Wylie, P. J. (April, 1996). Infrastructure and Canadian economic growth 1946–1991. *The Canadian Journal of Economics/Revue canadienne d'Economique, 29* [Speical Issue: Part 1], pp. S350–S355.

Yiu, C., Lu, X., Leung, M., & Jin, W. (2004). A longitudinal analysis on the relationship between construction output and GDP in Hong Kong. *Construction Management and Economics, 22* (4), 339–345.

Zheng, S., & Liu, H. (2004). Interaction among construction investment, other investment and gdp in China. *Tsinghua Science and Technology, 9*(2), 160–167.

Chapter 4
Trade and Economic Development

Abstract This chapter provides an overview of different perspectives on the connection between international trade and economic development. The chapter focuses on the debate between neoclassical trade theory and dependency theory. A number of contributions of international trade to economic development and basic macrolevel issues involved are illustrated. Besides, how governments manage trade policies and challenges faced by national policy-makers, including infrastructure development and renovation, are explored in this chapter.

Keywords International trade · Neoclassical trade theory · Dependency theory · Trade policies · Government institutions

4.1 Introduction

Developing countries are increasingly becoming concerned with international trade and market issues. The connection between international trade and economic development was first recognized in the literature on mercantilism in the sixteenth century. Since then, there have been different stances on the relationship. The first part of this chapter outlines these different perspectives and basic macrolevel issues involved in international trade in the developing countries. A number of contributions of international trade to economic development are illustrated. Given the importance of international trade in economic development, integrating into international market has become a critical issue for public policy in developing countries. Hence, how governments manage trade policy and challenges faced by national policy-makers are also explored in this chapter.

4.2 Trade Between Nations and Some General Features

Economists typically use the term "trade" to refer to the exchange of goods and services between individual consumers, between regions or between nations. Trade viewed in different ways has distinguishing features. Since this study focuses on a

G. Dang and L. Sui Pheng, *Infrastructure Investments in Developing Economies*,
DOI 10.1007/978-981-287-248-7_4

country's perspective on trade, the subject matter of this study concerns trade between nations or international trade. More specifically, international trade in this study is defined as all inward and outward movements across national borders of goods and services. One of the distinguishing features of international trade is the trade barriers that nations impose against trade (Markusen et al. 1995). Such trade barriers can range from traditional mechanisms of tariffs and quotas to "behind-the-border" issues, such as supporting infrastructure and governance. The rapid growth of international trade, especially in developing countries, has increased the significance of these trade barriers for export interests in these countries (Hoekman et al. 2002).

Globally, international trade has experienced high growth rates in recent decades. After World War II, the average annual growth rate of world merchandise exports in volume terms (i.e. in real terms, accounting for changes in prices and exchange rates) was 8 % per year over the 1950–1973 period. Over the 1990–1999 period, the rate was around 7 % per year. The rate over the 2000–2007 period was 6 % per year (World Trade Organization 2001, 2008). After being affected by the financial crisis that began at the end of 2008, the world merchandise exports in 2010 surged 14.5 % (World Trade Organization 2010). It is notable that trade between developing countries started to grow rapidly in the 1990s (Hoekman et al. 2002). In 2010, both exports and imports in real terms grew faster in developing countries than in developed ones. In 2010, the growth in the volume of exports in developed countries was 13 %, whereas the rate in developing countries was 17 %. Similarly, the 2010 imports in developed countries grew by 11 %, while that in developing countries was 18 % (World Trade Organization 2010).

International trade has become an important part of the overall economic activity of developing countries. Developing countries opened up their economies by implementing trade liberalization. They also increased their integration into the world economy through acceptance of international trade agreements (Markusen et al. 1995; Hoekman et al. 2002). Trade policy has become a critical element of development agenda in these economies (Hoekman et al. 2002). Nevertheless, the debate on the role of international trade and trade policy in developing countries has been characterized by different theoretical viewpoints as reviewed in the following sections.

4.3 International Trade and Economic Development

The relationship between international trade and economic development has long been a debatable topic, starting from mercantilism in the sixteenth century. Since the eighteenth century, the debate continues mainly between neoclassical trade theory and dependency theory (Hein 1992; Mullen et al. 2001; Zhang 2008).

Mercantilism was the main stream of economics of the sixteenth and seventeenth century. The main premise of mercantilism is that a nation can increase capital, which is necessary for the nation's prosperity, through positive balance of trade with other nations. Thus, mercantilists encourage governmental intervention that promotes exports and discourages imports. Mercantilists view international trade as a zero-sum game, in which benefits of one nation would be costs of another. Although the theory's arguments still have applications today, they began to fade in the late eighteenth century with widespread criticisms from classical economists, especially from Smith (1976, original work published 1776) (Zhang 2008).

The neoclassical trade theory considers free trade as an engine of economic growth. Starting from classical theory of absolute advantage by Smith (1976, original work published 1776), the basic principle of absolute advantage suggests that both countries trading with each other can gain from trade by selling to the other what they produce more cheaply than the other. Adam Smith's trade theory thus holds that trade is a positive-sum game. Refining the theory of absolute advantage, Ricardo's theory of comparative advantage (Ricardo 1971, original work published 1817) explained that absolute advantage was unnecessary for trade to occur. Even when both commodities may be produced more cheaply in one country than in another, trade of the commodities between the two countries can still occur. Because of differences in technology between countries, the relative efficiency among the two goods in one country is different from that in another country; thus, the former will gain by exporting the good which it produces relatively more efficient to the latter and importing the one which it has least absolute advantage from the latter. The theory hence holds that all participating countries share trade gains.

Different from the Ricardian trade theory, the Heckscher–Ohlin theory (Ohlin 1933) demonstrated that even when all countries have the same access to technology, costs of production between countries are still different; thus, participating countries can still enjoy comparative advantage. According to the theory, comparative advantage of a country is determined by its relative abundance or scarcity of factors of production (labour, capital and land) (Flam and Flanders 1991). Because of differences in resource endowments, a nation can produce goods that intensively use its relatively abundant resource cheaper than another, and vice versa; thus, trade occurs. The essential gains from international trade of a country therefore come from the reallocation of resources from one sector to another as the country increases specialization and concentrates investments in those sectors based on comparative advantage. On the global basis, international trade can help transfer the services of factors of production from countries where these factors are abundant to countries where they are scarce, leading to positive economic growth for both exporters and importers (Bhagwati 1990; Mullen et al. 2001; Katsioloudes and Hadjidakis 2007).

Although the Ricardian model and Heckscher–Ohlin theory are the foundation for the theory of international trade, their contributions only deal with given and static supplies of factors of production (Zhang 2008). More recently, the contemporary trade theory argues that even countries where factors are scarce can still enjoy comparative advantages through innovation. In these countries, firms have strong incentives to make efficient use of the scarce resources and be innovative, thus creating comparative advantage (Katsioloudes and Hadjidakis 2007). Through trade and foreign direct investment, firms can also achieve economies of scale and acquire new knowledge and technology from abroad, thus producing more efficiently and increasing overall productivity. These long-term effects of trade on economic growth have been the focus of new trade theory and new growth theory since the early 1970s (Zhang 2008). Overall, the important implication about the effects of trade on economic development in these theories is that through more efficient allocation of resources that enhances productivity, trade enables a nation to move beyond its production possibility frontiers, thereby acting as an engine of economic growth (Helpman and Krugman 1985; Grossman and Helpman 1991). Other developmental roles of exports and trade include increasing foreign exchange, creating employment and improving living standard (Mullen 1993; Awokuse 2007). On a global basis, other benefits brought by international trade include improvements in labour standards, incomes and fair trade for the developing countries (Mazur 2000; The Economist 2001).

The dependency theory, a neo-Marxist predecessor of the world-systems theory, by contrast, argues that exporting and importing will lead to slow economic growth for the developing countries. Marxist theory (Marx 1933, original work published 1867) holds that trade is not a positive-sum game, and capitalist countries benefit most from trade with poor countries through colonialism and imperialism (Pincus 1967). Trade, as viewed by the dependency theory, is actually a mechanism for the industrialized countries with multinational firms to extract surplus wealth from less developed countries (Ram 1985; Hein 1992). These multinational firms invest in less developed countries to produce exports but only to send the trade gains back to their home countries. By securing favourable terms of trade, the core countries (advanced industrial countries) transfer value from the peripheral countries (less developed countries) rather than reinvest in these countries (Galtung 1971; Dholakia and Dholakia 1984). Depending on foreign capital and markets is therefore considered as means to pave the way for multinational corporations to control less developed countries to benefit their home countries, thus stifling the development in less developed countries. From the theoretical perspective, the only way for less developed countries to develop is to become independent of foreign capital and goods (Hein 1992). Similarly, opponents of globalization mainly argued that globalization threatens state democracy and intensifies inequality for the poorer participants in international trade flows (Taylor 2002).

As a result, dependency and neoclassical trade theorists suggest extremely diverse external trade policies for developing countries, which have been a subject of much interest in the development literature as briefly reviewed earlier in Chap. 2 of this study. Neoclassical trade theorists with an outward-oriented perspective suggest export-led growth development strategies, whereas the dependency theorists with an inward-oriented perspective support the import substitution growth model. Outwardly oriented countries focus on trade liberalization (Krueger 1990). Inwardly oriented countries, by contrast, give priority to the domestic production of goods, which could otherwise be imported from the international market (Hein 1992). Import substitution regimes thus employ measures such as licensing procedures, quotas, tariffs or other barriers to restrict or prohibit imports, whereas export-oriented policies generally avoid all these restrictions to permit access to the international market at international prices, at least to exporters for their inputs, and use incentives to stimulate exports to the international market. It is worthwhile to mention that an outward-oriented strategy is not necessary to be completely free trade. There could still be elements of protection in an outward-oriented strategy as long as there is no bias of the incentive structure towards favouring the production of import substitutes (Krueger 1990; Thirlwall 2000).

After the Second World War, a widespread acceptance of import substitution policies as a means to stimulate economic growth developed among the developing countries, who had just obtained their independence from the colonial powers. Policy-makers of these newly independent countries were impressed by the historical experience of the more developed countries where high levels of import protection had contributed positively to industrialization in the nineteenth and twentieth century (Baldwin 2003). For example, in the 1970s, Latin American countries applied policies specifically suggested by the dependency theory, which imposed high levels of import duties and other import barriers, and maintained a hostile attitude towards foreign investment (Hein 1992). Import substitution policies actually worked quite well initially in these countries. However, in the 1990s, the drawbacks of such a policy approach became increasingly apparent. The permanent development was hampered by domestically high prices, and markets for consumer and industrial goods eventually became saturated. As a result, Latin American nations experienced stagnant growth with immense external debts, high rates of inflation, shortages of investment capital, as well as the increasingly downgrading of social and economic well-being of large segments of their population (Gereffi 1989; World Bank 2000).

In contrast to most of Latin American countries, East Asian countries, the first group of developing countries that shifted from an inward-oriented to an outward-oriented approach to development, experienced the fastest growth of exports and GDP in the 1970s (Hein 1992; Baldwin 2003). The shift to export-led growth policies, especially attempts to attract foreign investment in the 1960s and 1970s of these nations, has built up flexible, dynamic economies with high rates of growth (Hein 1992). The success of the Asian newly industrializing countries (NICs) has challenged the conventional thinking of scholars and policy-makers, opening new development possibilities for the developing countries. To varying degrees, a number of empirical studies across countries have also showed results in favour of outward-looking over inward-looking policies. Balassa (1971) and Little et al. (1970) compared import substitution policies across industries and countries and recommended developing countries to reduce the average levels of effective protection and the discrimination against exports. Bhagwati (1978) and Krueger (1978) studied changes in trade policies as well as other macroeconomic policies in specific countries. These studies concluded that it was the outward-oriented, instead of inward oriented, strategy that generated long-term high growth rate for developing countries. As a result, export-led growth strategy is considered the rational and efficient alternative to import substitution strategy for development (Awokuse 2007). Other studies attempted to draw correlation between some measures of openness, particularly value of exports, and economic growth or productivity growth. For example, Balassa (1978), Fosu (1990) and Krueger (1978) found a significant positive relationship between growth of exports and growth of GNP. A study by Michaely (1977) showed a positive association between per capita GNP and export share in GNP. Similarly, a number of empirical evidence have concluded the important effects of export performance, export expansion and openness to foreign markets on economic growth (Krueger 1980; Ram 1987; Moschos 1989; Islam 1990; Salvatore and Hatcher 1991; Sachs et al. 1995; Al-Yousif 1997; Dollar and Kraay 2002).

There is, however, some doubt about the causality between exports, imports and economic growth in the developing economies. An empirical study by Dodaro (1993) showed a weak support for the view that export growth would promote GDP growth in developing countries. Palley (2003) was sceptical of the sustainability of export-led growth, although the focus was on global growth rather than growth of individual countries. According to the study, export-led development may work for a few countries, but it is a zero-sum game on a global basis. In other studies, Bhagwati (1988) and Krugman (1992) argued that there could be reverse causal flow from economic growth to exports growth. Productivity gains through increases in domestic levels of skilled labour and technology can lead to export expansion. Similarly, an important review by Rodriguez and Rodrik (2000) demonstrated that successful export performance could be a result of successful development rather than cause, although the study only focused on the effects of trade barriers on growth rather than on more general measures of openness. Also, there could be a bidirectional causality between exports and economic growth (Dutt and Ghosh 1996; Bekõ 2003). As explained by Dutt and Ghosh (1996), in the initial stages of development of some countries, economic growth stimulates exports, but along the way exports will lead to further economic growth. Moreover, Dutt and Ghosh (1996) believed that the causality structure depends on characteristics of a specific economy; thus, it would be inappropriate to generalize the structure.

Despite the ongoing debate, the empirical evidence available indicates that no nation has ever maintained high economic growth over the long term without trade (Markusen et al. 1995). Empirical evidence shows that those countries that deepened their integration with the global economy have achieved growth much higher than those developing countries that did not or have slowly integrated into global markets (Hoekman et al. 2002; Newfarmer et al. 2002; Samli 2002). Opening to international competition is considered the starting point for developing countries (Dominguez et al. 1993). The important issue for developing countries is thus not about whether to trade but should be more about how to trade in order to secure improvement in economic performance (Thirlwall 2000; Taylor 2002). This enhances the significant role of trade facilitation policies that are critical in supporting international trade (Rodrik 2002). Trade facilitation is broadly defined to consist of all factors that support a country's capacity and effectiveness to create and maintain a trade-friendly environment, including border barriers to trade (such as port efficiency and customs administration) and "behind the border" (such as quality of infrastructure and regulatory environment) (Wilson et al. 2005).

4.4 International Market Access Issues and Trade Policies for Developing Countries

The integration of developing countries into the international trading system has been extensively analysed in the past 20 years. The variables in most studies on macrostrategy at the national level include trade barriers, institutions that support the participation of national firms in international markets, trade liberalization and competitive advantage (Kilbourne 2004). The first three variables have long been reviewed and analysed in a number of studies within international trade literature. The results of these studies were used to develop negotiation as well as incentives by the international community and developing countries aimed at more effective integration of developing countries into the international economy (Michalopoulos 1999b; Hoekman 2002). These variables are considered to have a clear link with trade policies and market access issues for developing countries. Since trade barriers and trade protection are considered harmful for effective integration into the multilateral trading systems, most studies highlight the importance of opening up the trade system to competing traders, lowering the use of tariff and non-tariff barriers, removing exchange rate distortions, and the possible introduction of measures to reduce the impact of continuing distortions (Michalopoulos 1999b; Martin 2001). Integrating into the multilateral trading systems, including the World Trade Organization (WTO), also requires the developing countries to implement trading agreements governed by the WTO such as the General Agreement on Tariffs and Trade (GATT) for trade in goods, the General Agreement for Trade in Services (GATS) for trade in services, as well as other aspects of international exchange of goods and services, such as trade-related intellectual property rights (TRIPS), sanitary and phytosanitary standards (SPS), government procurement and so on (Michalopoulos 1999a; Hoekman et al. 2002).

The main challenges for developing countries are the constraints in their capacity of broad, market-based institutions and those which are more specifically trade related, such as the financial sector, as well as customs and trade facilitation. Developing countries are therefore recommended to combine reducing trade barriers with other policies that are conducive to improving the investment environment and macroeconomic stability, and strengthening financial, regulatory and institutional capacity with technical and financial assistance from bilateral donors and international organizations (Rodrik 1999; Wilson 2001; Prowse 2002). Upgrading transport, communication and other basic infrastructure services and building capacity to meet global regulatory requirements and standards are also important to the developing countries in order to expand trade opportunities and to effectively integrate them into the world trade system (Michalopoulos 1999a, b; Hoekman 2002).

The macrostrategies at the national level, including the trade agreements and practices, recommended above were generally based on the comparative advantage doctrine (Samli and Jacobs 1995). However, more recently, it is argued that comparative advantage may not provide the complete answer for the development of an efficient export sector; a more coherent strategy may lie in competitive advantage (Dominguez et al. 1993; Lowinger et al. 1995). The competitive advantage model was proposed and advanced by Porter (1990, 1998a, b, 2000). Porter's works were based on the premise that the national competitive environment in which firms operate can considerably influence the international competitive performance of firms. The competitiveness of nations viewed by Porter (1990, 1998a, b, 2000) is the ability of a nation to create and maintain such an environment that shapes highly competitive industries and firms. Firms built in such a national environment can be more successful when they export or engage in outward foreign direct investment (Rugman and D'Cruz 1993). According to Porter (1990, 1998a, b, 2000), elements of a national environment that play important roles in firms' competitive performance within the world market consist of the following:

- Factor conditions (i.e. the nation's position in factors of production, especially advanced factors of production such as skilled labour, capital and infrastructure);
- Demand conditions (i.e. sophisticated and demanding customers in the domestic market);
- Firm strategy, structure and rivalry (i.e. the characteristics of the business sector as conditions for organization of companies, and the nature of domestic rivalry); and
- Related and supporting industries (i.e. the importance of clustering industries and spillover benefits of an industry's investments in advanced factors of production within these clusters).

Porter (1990) integrated the four primary conditions into a diamond-shaped diagram with two exogenous parameters, namely government and chance. The four contextual variables operate interdependently, and their strength of interaction determines the extent to which a national environment is conducive to international success. A nation is therefore viewed as a set of contextual variables which can hinder or support the competitive performance of firms and industries (Grant 1991). As a result, the macrostrategy through which government can have the most direct effect in the competitive development process involves policies that support research and development, investments in human resource development, and infrastructure development and renovation (Dominguez et al. 1993; Samli and Jacobs 1995).

While Porter's diamond framework at the country level has become the emphasis within the management school, it has been ignored by the economics school. Recent studies within the economic literature argue that Porter's diamond framework is not about trade; it cannot explain the pattern of trade and gains from trade and thus cannot replace the theory of comparative advantage (Smit 2010). Moreover, Porter's explanation of international competitiveness of countries has not been undergone rigorous testing (Hill 2009). Nevertheless, Porter's diamond framework has offered an unprecedentedly broad range of national-level sources of competitive advantage and brought them together in a comprehensive model for formulating government strategy (Grant 1991; Rugman and D'Cruz 1993;

Moon et al. 1998; Peng 2009). Moreover, within the domain of international economics, the use of Porter's diamond framework has been discussed and considered as a good starting paradigm for greater optimization of national efforts to achieve success in international markets (Samli and Jacobs 1995; Moon et al. 1998; Smit 2010).

The importance of developing countries in the world economy is increasing. However, only a few of the higher income developing countries (developing countries based on WTO's definition with high GDP per capita, e.g. Singapore and South Korea) have impressively integrated into the multilateral trading system. The progress of many other developing countries has been slower. There were not much improvements in the conditions affecting access to international markets in these countries (Michalopoulos 1999b; Hoekman 2002). Although assistance from the international community is important for the progress, these developing countries' own policies are even more crucial. In the context of growing complexity and interdependence of foreign trade policies, designing and sustaining supporting policies are challenging tasks (Hoekman 2002; Jordana and Ramió 2003). The following section will examine the role of institutional structure in the formulation of foreign trade policy at the national level.

4.5 Trade Policy Institutions

Using international trade as a vehicle for a country's development strategy requires the country to open up and liberalize trade, which represents a change of direction in long-term development strategy and involves a new leading role for trade policy. Engagement with the multilateral trading systems also makes trade policy far more complex, which creates pressures to reinforce the trade policy process and the strategic management capacity in the main organization responsible for trade policy making (Rodrik 2002; Jordana and Ramió 2003). The related issues as discussed above place the focus of trade policy on creating incentives for international trade growth by reducing border barriers to trade (tariff and non-tariff barriers), strengthening the public institutions needed for efficient customs, improving legal and regulatory environment and creating measures to enhance the competitiveness of national firms in international markets. Institutional structure and the formulation of international trade policy at the national level are considered critical elements for an efficient trade policy process (Hoekman 2002; Devlin and Estevadeordal 2003; Lecomte 2003; Sinha 2007). As a result, there is a growing need for governments in developing countries where institutional performance remains weak to introduce further changes into their existing institutional models.

Responding to the above-mentioned challenges, there have been efforts in many developing countries to renew and adapt themselves to the demands of the new international environment. Transformation of institutional structure is often characterized by giving greater power to the core ministry dealing with trade (Jordana and Ramió 2003). For example, the power and status of India's Ministry of

Commerce has been radically enhanced recently, in part to attempt to further export its products and services aggressively and to engage the WTO more actively. The growing complexity of trade policy also necessitates the involvement of, or the integration with, other government ministries in the policy process such as the Ministry of Foreign Affairs, Ministry of Industry or Ministry of Agriculture (Sinha 2007). The inclusion of all actors, including the relevant ministries, trade promotion and regulatory bodies, and non-governmental players from the business sector, think tanks, independent research centres, leading universities and other civil society organizations is also considered part of the institutional transformation (Lecomte 2003). However, introducing changes into an organizational model is not a trivial task. Moreover, different conditions in each country require a specific institutional design. In many developing countries, performance has not been improved satisfactorily. Trade policy formulation and implementation in these countries are still dominated by governmental and intergovernmental agencies with overlapping functions and deficient coordination (Jordana and Ramió 2003; Jerome 2005). Further changes in the existing institutional structure and improvements in institutional capacity are thus vital if trade liberalization is to serve as an engine of growth.

4.6 Summary

The relationship between international trade and economic development varies in the literature. International trade is considered an engine of economic growth in the neoclassical economic trade theory. This conflicts with the dependency theory, in which international trade is viewed as a hindrance to economic development. A summary of the major theoretical insights of the relationship is given in Fig. 4.1.

However, focusing more on the empirical perspective, a number of studies have shown a positive association between export expansion and openness to foreign markets, and economic development. Although there is still some doubt about the causality of the relationship, international trade between nations is believed to play a crucial role in the economic development process. It is international trade that links a country with its international market, and drives domestic resource allocation to be utilized effectively. As a result, what is needed for economic development in most developing countries is to effectively engage in international trade.

Effective integration into the international market or the world trading system requires a liberal trade policy using traditional policy instruments such as tariffs, quotas, export subsidies and taxes, complemented by trade-related institutions to facilitate trade and attain national objectives. A more recent suggested trade agenda covers a broader focus on such policies as those involving research and development, human resources development, as well as infrastructure renovation and development.

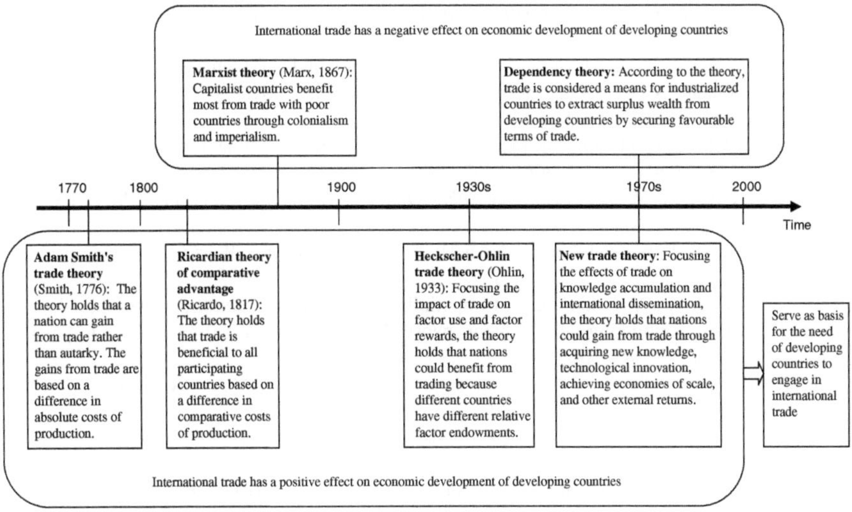

Fig. 4.1 Timeline of major theoretical viewpoints on the role of international trade in economic development

References

Al-Yousif, Y. K. (1997). Exports and economic growth: Some empirical evidence from the Arab Gulf countries. *Applied Economics, 29*(6), 693–697.

Awokuse, T. O. (2007). Causality between exports, imports, and economic growth: Evidence from transition economies. *Economics Letters, 94*(3), 389–395.

Balassa, B. A. (1971). *The structure of protection in developing countries*. Baltimore: The Johns Hopkins Press.

Balassa, B. A. (1978). Exports and economic growth: Further evidence. *Journal of Development Economics, 5*(2), 181–189.

Baldwin, R. E. (2003). *Openness and growth: What's the empirical relationship?* (NBER Working Paper No. 9578). Cambridge, MA: National Bureau of Economic Research.

Bekõ, J. (2003). Causality analysis of exports and economic growth: Aggregate and sectoral results for Slovenia. *Eastern European Economics, 41*(6), 70–92.

Bhagwati, J. N. (1978). *Foreign trade regimes and economic development: Anatomy and consequences of exchange control regimes*. Cambridge, MA: Ballinger.

Bhagwati, J. N. (1988). *Protectionism*. Cambridge, MA: MIT Press.

Bhagwati, J. N. (1990). Export-promoting trade strategy: Issues and evidence. In C. Milner (Ed.), *Export promotion strategies: Theory and evidence from developing countries* (pp. 11–39). London: Harvester Wheatsheaf.

Devlin, R., & Estevadeordal, A. (2003). *Bridges for development: Policies and institutions for trade and integration*. Washington, D.C.: Inter-American Development Bank.

Dholakia, N., & Dholakia, R. R. (1984). Missing links: Marketing and the newer theories of development. In G. S. Kindra (Ed.), *Marketing in developing countries* (pp. 57–75). New York: St. Martin Press.

Dodaro, S. (1993). Exports and growth: A reconsideration of causality. *Journal of Developing Areas, 27*(2), 227–244.

Dollar, D., & Kraay, A. (2002). Growth is good for the poor. *Journal of Economic Growth, 7*(3), 195–225.
Dominguez, L., Rose, P., & Sequeira, C. (1993). Environmental and managerial factors in export-led development: An explorative test of Porter's competitive advantage of nations. *Journal of Macromarketing, 13*(2), 5–21.
Dutt, S. D., & Ghosh, D. (1996). The export growth-economic growth nexus: A causality analysis. *Journal of Developing Areas, 30*(2), 167–182.
Flam, H., & Flanders, M. J. (1991). *Heckscher-Ohlin trade theory*. Cambridge, MA: The MIT Press.
Fosu, A. K. (1990). Exports and economic growth: The African case. *World Development, 18*(6), 831–835.
Galtung, J. (1971). A structural theory of imperialism. *Journal of Peace Research, 8*(2), 81–117.
Gereffi, G. (1989). Rethinking development theory: Insights from East Asia and Latin America. *Sociological Forum, 4*(4), 505–533.
Grant, R. M. (1991). Porter's 'Competitive advantage of nations': An assessment. *Strategic Management Journal, 12*(7), 535–548.
Grossman, G. M., & Helpman, E. (1991). *Innovation and growth in the global economy*. Cambridge, MA: MIT Press.
Hein, S. (1992). Trade strategy and the dependency hypothesis: A comparison of policy, foreign investment, and economic growth in Latin America and East Asia. *Economic Development and Cultural Change, 40*(3), 495–521.
Helpman, E., & Krugman, P. R. (1985). *Market structure and foreign trade: Increasing returns, imperfect competition, and the international economy*. Cambridge, MA: MIT Press.
Hill, C. W. L. (2009). *International business: Competing in the global marketplace* (7th ed.). Boston: McGraw-Hill/Irwin.
Hoekman, B. (2002). Strengthening the global trade architecture for development: The post Doha agenda. *World Trade Review, 1*(01), 23–45.
Hoekman, B., Mattoo, A., & English, P. (Eds.). (2002). *Development, trade, and the WTO: A handbook*. Washington, D.C.: World Bank Publications.
Islam, N. (1990). *Horticultural exports of developing countries: Past performances, future prospects, and policy issues* (Research Report No. 80). Washington, D.C.: International Food Policy Research Institute.
Jerome, A. (2005). *Institutional framework and the process of trade policy making in African: The case of Nigeria*. Paper presented at the international conference in African economic research institutions and policy development: Opportunities and challenges. Retrieved from http://www.idrc.ca/uploads/user-S/11062328591Institutional_Framework_and_the_Process_of_Trade_Policy.pdf.
Jordana, J., & Ramió, C. (2003). Trade policy institutions: Comparative analysis. In R. Devlin & A. Estevadeordal (Eds.), *Bridges for development: Policies and institutions for trade and integration* (pp. 179–201). Washington, D.C.: Inter-American Development Bank.
Katsioloudes, M. I., & Hadjidakis, S. (2007). *International business: A global perspective*. Burlington, MA: Butterworth Heinemann.
Kilbourne, W. E. (2004). Globalization and development: An expanded macromarketing view. *Journal of Macromarketing, 24*(2), 122–135.
Krueger, A. O. (1978). *Liberalization attempts and consequences*. Cambridge, MA: Ballinger.
Krueger, A. O. (1980). Trade policy as an input to development. *The American Economic Review, 70*(2), 288–292.
Krueger, A. O. (1990). *Perspectives on trade and development*. New York: Harvester Wheatsheaf.
Krugman, P. R. (1992). Import protection as export promotion: International competition in the presence of oligopoly and economies of scale. In G. M. Grossman (Ed.), *Imperfect competition and international trade* (pp. 75–86). Cambridge, MA: MIT Press.
Lecomte, H. B. S. (2003). *Building capacity to trade: What are the priorities?* (Working Paper No. 223). Paris: OECD.

Little, I. M. D., Scitovsky, T., & Scott, M. F. (1970). *Industry and trade in some developing countries: A comparative study*. New York: Oxford University Press.
Lowinger, T. C., Lal, A. K., & Arunthanes, W. (1995). Foreign access to the Japanese market: Structural and economy-wide factors. *Journal of Macromarketing, 15*(2), 50–65.
Markusen, J. R., Melvin, J. R., Maskus, K. E., & Kaempfer, W. (1995). *International trade: Theory and evidence*. Singapore: McGraw-Hill.
Martin, W. (2001). *Trade policy reform in the East Asian transition economies* (Policy Research Working Paper No. 2535). Washington, D.C.: The World Bank.
Marx, K. (1933). *Capital*. London: J. M. Dent.
Mazur, J. (2000). Labor's new internationalism. *Foreign Affairs, 79*(1), 79–93.
Michaely, M. (1977). Exports and growth: An empirical investigation. *Journal of Development Economics, 4*(1), 49–53.
Michalopoulos, C. (1999a). *The integration of transition economies into the world trading system* (Policy Research Working Paper No. 2182). Washington, D.C.: World Bank.
Michalopoulos, C. (1999b). *Trade policy and market access issues for developing countries: Implications for the millennium round* (World Bank Policy Research Working Paper No. 2214). Washington D.C.: World Bank.
Moon, H. C., Rugman, A. M., & Verbeke, A. (1998). A generalized double diamond approach to the global competitiveness of Korea and Singapore. *International Business Review, 7*(2), 135–150.
Moschos, D. (1989). Export expansion, growth and the level of economic development: An empirical analysis. *Journal of Development Economics, 30*(1), 93–102.
Mullen, M. R. (1993). The effects of exporting and importing on two dimensions of economic development: An empirical analysis. *Journal of Macromarketing, 13*(1), 3–19.
Mullen, M. R., Beller, E., Remsa, J., & Cooper, D. (2001). The effects of international trade on economic growth and meeting basic human needs. *Journal of Global Marketing, 15*(1), 31–55.
Newfarmer, R., Timmer, H., Varoudakis, A., Mattoo, A., Fink, C., Maskus, K., et al. (2002). *Global economic prospects and the developing countries 2002: Making trade work for the world's poor* (Publication No. 23590). Washington, D.C.: World Bank.
Ohlin, B. G. (1933). *Interregional and international trade*. Cambridge: Harvard University Press.
Palley, T. I. (2003). Export-led growth: Evidence of developing country crowding-out. In P. Arestis, M. Baddeley, & J. S. L. McCombie (Eds.), *Globalisation, regionalism and economic activity* (pp. 175–197). Cheltenham: Edward Elgar.
Peng, M. W. (2009). *Global business*. Mason, Ohio: South-Western Cengage Learning.
Pincus, J. A. (1967). *Trade, aid and development: The rich and poor nations*. New York: McGraw-Hill.
Porter, M. E. (1990). *The competitive advantage of nations*. New York: Free Press.
Porter, M. E. (1998a). Clusters and the new economics of competition. *Harvard Business Review, 76*(6), 77–90.
Porter, M. E. (1998b). *The competitive advantage of nations: With a new introduction*. New York: Free Press.
Porter, M. E. (2000). Location, competition, and economic development: Local clusters in a global economy. *Economic Development Quarterly, 14*(1), 15–35.
Prowse, S. (2002). The role of international and national agencies in trade–related capacity building. *World Economy, 25*(9), 1235–1261.
Ram, R. (1985). Exports and economic growth: Some additional evidence. *Economic Development and Cultural Change, 33*(2), 415–425.
Ram, R. (1987). Exports and economic growth in developing countries: Evidence from time-series and cross-section data. *Economic Development and Cultural Change, 36*(1), 51–72.
Ricardo, D. (1971). *On the principles of political economy, and taxation*. Harmondsworth: Penguin.
Rodriguez, F., & Rodrik, D. (2000). Trade policy and economic growth: A skeptic's guide to the cross-national evidence. *NBER Macroeconomics Annual, 15*, 261–325.

Rodrik, D. (1999). *The new global economy and developing countries: Making openness work.* Washington, D.C.: Overseas Development Council.
Rodrik, D. (2002). Trade policy reform as institutional reform. In B. Hoekman, A. Mattoo, & P. English (Eds.), *Development, trade, and the WTO: A handbook* (pp. 3–10). Washington, D.C.: The World Bank.
Rugman, A. M., & D'Cruz, J. R. (1993). The "double diamond" model of international competitiveness: The Canadian experience. *Management International Review, 33*(2), 17–39.
Sachs, J. D., Warner, A., Åslund, A., & Fischer, S. (1995). Economic reform and the process of global integration. *Brookings Papers on Economic Activity, 1995*(1), 1–118.
Salvatore, D., & Hatcher, T. (1991). Inward oriented and outward oriented trade strategies. *Journal of Development Studies, 27*(3), 7–25.
Samli, A. C. (2002). *In search of an equitable, sustainable globalization: The bittersweet dilemma.* Westport, CT: Quorum Books.
Samli, A. C., & Jacobs, L. (1995). Achieving congruence between macro and micro generic strategies: A framework to create international competitive advantage. *Journal of Macromarketing, 15*(2), 23–32.
Sinha, A. (2007). Global linkages and domestic politics: Trade reform and institution building in india in comparative perspective. *Comparative Political Studies, 40*(10), 1183–1210.
Smit, A. J. (2010). The competitive advantage of nations—Is Porter's diamond framework a new theory that explains the international competitiveness of countries. *Southern African Business Review, 14*(1), 105–130. Retrieved from http://www.unisa.ac.za/contents/faculties/service_dept/docs/SABVIEW14_1_chap5.pdf.
Smith, A. (1976). *An inquiry into the nature and causes of the wealth of nations.* Oxford: Clarendon Press.
Taylor, A. M. (2002). *Globalization, trade, and development: Some lessons from history* (NBER Working Paper No. w9326). Cambridge, MA: National Bureau of Economic Research.
The Economist. (2001). Globalisation and its critics. *The Economist, 360*(8241), 3–5.
Thirlwall, A. P. (2000). *Trade, trade liberalisation and economic growth: Theory and evidence* (Economic Research Paper No. 63). Côte d'Ivoire: African Development Bank.
Wilson, J. S. (2001). *Bridging the standards divide: Recommendations for reform from a development perspective* (Working Paper No. 27824). Washington D.C.: World Bank.
Wilson, J. S., Mann, C. L., & Otsuki, T. (2005). Assessing the benefits of trade facilitation: A global perspective. *The World Economy, 28*(6), 841–871.
World Bank. (2000). *Entering the 21st century—World development report 1999/2000.* New York: Oxford University Press.
World Trade Organization. (2001). *International trade statistics 2001.*
World Trade Organization. (2008). *Trade in a globalizing world* (World Trade Report 2008).
World Trade Organization. (2010). *Trade in natural resources* (World Trade Report).
Zhang, W.-B. (2008). *International trade theory: Capital, knowledge, economic structure, money, and prices over time.* Germany: Springer.

Chapter 5
Public Investments in Infrastructure for Trade and Economic Development

Abstract This chapter presents a discussion on the intersection between trade, infrastructure construction and economic development. The chapter focuses on the theory of coordination failure and the role of government in infrastructure development for trade and economic growth. Governments in developing countries are advised to increase investments in infrastructure to stimulate international trade and economic growth. However, there are concerns about the low efficiency of public investments in infrastructure construction. The chapter discusses these concerns and the knowledge gap. The chapter also reviews the literature to identify major issues that could affect the efficiency of government-funded infrastructure in supporting trade and economic growth.

Keywords Theory of coordination failure · Trade · Infrastructure construction · Economic development · Role of government

5.1 Introduction

A review of the literature on the role of infrastructure construction as well as the role of international trade in economic development in previous chapters has illuminated several intriguing aspects of sectoral linkages in the dynamic process of economic development. Governments in developing countries are advised to increase investments in infrastructure to stimulate international trade and economic growth. However, there are concerns about the low efficiency of public investments in infrastructure construction. These concerns are discussed in this chapter.

G. Dang and L. Sui Pheng, *Infrastructure Investments in Developing Economies*,
DOI 10.1007/978-981-287-248-7_5

5.2 Theory of Coordination Failure and Role of Government

As reviewed in Chap. 2, one strand of modern theory of economic development approaches underdevelopment as a coordination problem. The logic of coordination failures contends that when complementaries exist, firms' inability to coordinate their behaviours leads the economy to an inefficient state (Matsuyama 1997; Todaro and Smith 2003). Models of coordination failures demonstrate that even market mechanism cannot solve the problem "because of difficulties of coordination, sometimes because people hold different expectations, sometimes because everyone is better off waiting for someone else to make the first move" (Todaro and Smith 2003, p. 150). The first researcher to raise some of the basic coordination issues was Rosenstein-Rodan (1943). Rosenstein-Rodan (1943) found the sources of coordination failures by focusing on complementaries in production decisions made by firms in the process of industrialization. Rosenstein-Rodan (1943) and later models based on Rosenstein-Rodan (1943), such as Murphy et al. (1989) and Krugman (1995) called for a "big push" industrialization, which is a synchronized expansion of industries coordinated by central government planning.

Studies in development economics such as those completed by Matsuyama (1997) and Hoff (2000) described the economy as a complex system of activities with interrelated production processes. "The development of a sophisticated economic system requires a high degree of coordination among these diverse activities, performed by a diverse set of agents, each of whom may possess the unique knowledge and technical expertise concerning these activities" (Matsuyama 1997, p. 142). The major part of the coordination problem is to figure out which combination of activities could bring about a better outcome for the economy as a whole (Matsuyama 1997).

The presence of coordination failures that may arise in the presence of complementaries between these activities highlights potential policies for interventions that remove the coordination problems and make the most efficient use of the available resources, thus achieving economic development. However, the role of government in improving the way to coordinate economic activities is not consistent in the literature on coordination problems. Recent works undermined government intervention to solve coordination problems. A bad policy choice could also be one of the important sources of a bad equilibrium. In many developing countries, the government appears to be inefficient and is at the root of most important poverty traps. The government policies are thus not superior to the market mechanism in coordinating economic activities (Glăvan 2008). Nonetheless, it is the focus of this study that the government cannot abandon its role as coordinator and facilitator of overall economic development, especially in the developing countries where market failure is high as well. Government interventions are important to ensure that in the long-run the progress that a country makes is targeted at social and economic priorities. From that perspective, this study will discuss the role of government in coordinating infrastructure development and trade growth in the following sections.

5.3 Public Investments in Infrastructure and Knowledge Gap

As reviewed earlier in Chap. 4 of this study, the literature on the relationship between international trade and economic development has been shaped by different theoretical views (Low and Dang 2012). However, empirical studies have shown that developing countries should focus on how to trade in order to secure improvement in economic performance rather than on whether to trade (Thirlwall 2000; Taylor 2002). Measures to stimulate international trade growth underpinned by the comparative advantage doctrine (Smith 1976; Ricardo 1971; Ohlin 1933) emphasize the importance of issues relating to trade barriers, institutions and trade liberalization. Meanwhile, the recent work of the competitive advantage doctrine (Porter 1990, 1998a, 2000) has expanded the focus to the competitiveness of nations. The model views a nation as a set of contextual variables which can hinder or support the competitive performance of firms and industries. One of the main contextual variables that hinders or supports the competitive performance of firms and industries is infrastructure development.

There have been a number of empirical studies on the impact of infrastructure in general and physical distribution infrastructure in particular on trade and economic growth in the developing countries as briefly discussed earlier in this study. In these studies, infrastructure is regarded as one of the determinants of international competitiveness, which directly impact the ability of countries to engage in international trade, and to compete for foreign direct investments (FDI). The remarkable growth of a number of developing countries in Asia, including emerging East Asian economies, in recent decades has been attributable much to the expansion of international trade, which in turn has been supported by infrastructure development (APEC Economic Committee 1997; Kohsaka 2006; Brooks 2008). However, their levels of infrastructure development still lag far behind that of the industrialized economies. In Africa, the level of infrastructure development is worse. Morrissey and Rudaheranwa (1998) and Nwankwo (2000) found that many African countries have not experienced the expected increase in export earnings although these countries have followed the policy of trade liberalization. Among a number of reasons for the poor performance of the export sector, the lack of proper and adequate infrastructure systems is one of the major concerns. Under-performance of infrastructure systems including telecommunications, energy and transport, as well as logistics infrastructure in many African countries have significantly increased investment and transaction costs, leading to an increase in entry barriers, as well as a reduction in competitiveness and access to domestic and international markets. Similarly, Peters (1992), Hulten (1996) and the Economic Research Center (2001) found that an adequate and reliable infrastructure system can promote trade, expand market through cost and time savings in transportation and telecommunications. Evaluating the importance of physical distribution and logistics infrastructure system in the future growth of horticultural exports of developing countries, Islam (1990) also maintained that improvement in transportation and other infrastructure for export is an important factor influencing their relative competitive positions in export markets, and thus their export performance. Furthermore, findings in an empirical study by Kumar (2001) showed that the availability of good quality physical infrastructure could improve the investment climate for FDI, especially export-oriented FDI. Apart from increased trade and foreign capital flows, the impact of infrastructure on trade can also be found in enhanced knowledge spillovers and technology transfer (Venables 2004). The lack of infrastructure in India was estimated to hinder the country's economic growth by 1.5 to 2 % per year (World Economic Forum 2007). Similarly, less investment in infrastructure was considered to be attributable to the slow economic growth of countries in Latin America (Fay and Morrison 2008).

The poor quality and inadequacy of infrastructure become one of the major development challenges in many developing countries. Considerable financing is required to meet large-scale infrastructure expansions (Fay and Yepes 2003; Davis 2008; Flanagan and Jewell 2009; Arnold 2011). However, investment in infrastructure is not directly seen to be feasible from individual private investors' point of view. This is because an infrastructural construction project is often capital intensive and is more conspicuous of its public goods characteristics. For example, a new transportation facility could bring about the improvement in the performance of all individual firms. However, it is not in the interests of any individual firm to address such issues. Moreover, the high costs of investment may render the private returns of the project unattractive. Meanwhile, the external effects of the consumption of such goods by the whole community go beyond the benefits to individual investors. In other words, the social rate of returns on such investments is higher than purely the private rate. In the event of market failure, it is the responsibility of the government to provide the investment goods for the economy as a whole. As a result, governments own and finance nearly all infrastructure projects worldwide. Globally, government investments accounted for 78 % of total investments in infrastructure construction from 1994 to 2003 (Estache 2006; Kenny 2007).

Since there is a huge need for additional investment, the literature is dominated by studies focusing on challenges in project financing for infrastructure development, including constraints in public expenditure and private financing (Fay and Yepes 2003; Davis 2008; Mulu and Smith 2008; Smith and Gannon 2008; Hwang et al. 2012). There are studies, however, suggesting a negative relationship between infrastructure investments and economic growth. A study of the US data by Kocherlakota and Yi (1996) illustrated that infrastructure does not permanently raise economic growth rate. Devarajan et al. (1996) indicated that infrastructure in developing countries actually had negative impacts on economic growth if infrastructure is oversupplied relative to the economic scale. Flyvbjerg (2007) and Flyvbjerg et al. (2002) when studying 258 transportation infrastructure projects around the world, the largest database of its kind thus far, indicated that cost overruns, benefit shortfalls and waste were found in most infrastructure projects. The low efficiency and quality of public investments in infrastructure found in these studies suggest that there are important issues beyond the difficulties in project financing for infrastructure development. A focus on financing issues thus does not provide a comprehensive answer to the infrastructure development problems in developing countries. Some studies focus on the planning and policy-making aspect of infrastructure development (Hossain 2000; Flyvbjerg 2007; Mustajab 2009; Marshall 2011). However, little is known about how the entire planning and implementation of infrastructure can be improved to help governments in developing countries cut down costs and waste, thus enhancing the efficiency of public investments in infrastructure, and meeting development goals. The knowledge gap requires further study on the way infrastructure is planned and built to support trade and economic development.

5.4 Infrastructure Planning and Implementing in Developing Countries

5.4.1 Problems in Infrastructure Planning

The connection between infrastructure construction and economic development reviewed earlier suggests that infrastructure development is one of the most integral parts in government planning and policy making. Infrastructure investments continue to be planned, decided and financed at the national level. Policy makers formulate decisions on infrastructure development based on infrastructure planning outcomes. However, there are a number of issues with the quality of these outcomes, including the absence of an adequate problem analysis, lack of alternatives, ambiguities about the effects of improved infrastructure on the development of a wider area, inadequate research of the interaction across infrastructure sectors, and underestimated costs and overestimated benefits (Estache and Fay 2007; Flyvbjerg 2007; Priemus 2010). The poor quality of infrastructure planning outcomes would thus result in poorly selected infrastructure investments and bad policy choices, which subsequently have a wider effect on the economy. Although planning tools have recently been developed (Schweikert and Chinowsky 2012; World Economic Forum 2012), there are deficiencies in the capacities required for using these tools. Tackling deficiencies in planning and policy-making capacities of governments can therefore play a crucial role in determining the efficiency of public investments in infrastructure in developing countries. From the literature review on government planning, especially in the area of infrastructure development, the following problems can be found attributed to deficiencies in the infrastructure planning and policy-making process.

5.4.1.1 Capacity for Estimation and Monitoring of Rates of Return of Projects

Unexpected infrastructure planning outcomes can first be explained by the lack of capacity for estimation and monitoring of rates of return of projects, including limitations of forecasting methods and appraisal techniques; inadequate data; inherent problems in predicting the future and monetizing external and indirect effects; lack of experienced forecasters; lack of quality checks on planning outcomes; and inadequacy in routine ex post analysis and external audits on whether policies and projects meet objectives (Short and Kopp 2005; Flyvbjerg 2007; Collier and Venables 2008; Todaro and Smith 2009).

5.4.1.2 Politicized Decision-Making

Besides these technical factors, failures of the planning process could be explained by the political factor (Todaro and Smith 2003; Flyvbjerg 2007; Guasch et al. 2007; Collier and Venables 2008). Political leaders and government bureaucrats can use investments in infrastructure construction as a tool for securing political positions or competing for scarce funds. In order to get approval and funding, investments in infrastructure projects backed by the political affiliations of the government would deliberately be proposed with overestimated benefits and underestimated costs. Risks in these projects would also be overlooked. Lack of commitment of political leaders and government bureaucrats to national goals as opposed to personal interests could therefore make infrastructure planning and decision-making politicized rather than rationalized. Empirical evidences from 69 developing countries over a 20-year time period indicated a statistically significant negative relationship between the share of central government investments on transport and communications, and per-capita GDP. The negative relationship was explained by political factors in decision-making (Devarajan and Swaroop 1993). In addition, powerful groups with vested interest can create pressure to affect the planning that serves their own interests (Todaro and Smith 2003; Henisz and Zelner 2006).

5.4.1.3 Transparency and Accountability

Another problem in infrastructure planning and policy making relates to the lack of transparency and accountability. These processes are rarely fully transparent to the public. Forecasting methods, criteria of project selection and the determination of planning objectives are often not made available for consultation with the interested or affected individuals or groups. Independent peer reviews and quality checks on forecasts and planning outcomes by independent review bodies and the scientific and professional community are not sufficient. Lack of penal systems to enforce penalties on those that deliberately and consistently produce deceptive forecasts is also attributable to the accountability problem (Short and Kopp 2005; Flyvbjerg 2007).

5.4.1.4 Institutional Weaknesses in Planning

Since planning and policy making for infrastructure is a multiactor process, nurturing of an innovative institutional capacity that coordinates efforts and resources is considered one of the determinants of infrastructure development (Mody 1997).

However, there are a number of institutional weaknesses of the planning processes of most developing countries, including the poor communication between the planning agency and the day-to-day decision-making machinery of government; intersectoral rivalries; lack of interaction between political leaders and planners with non-governmental actors; incompetent and unqualified civil servants; as well as complicated and bureaucratic administrative systems (Todaro and Smith 2003).

5.4.2 Problems in the Implementation of Infrastructure Plans

While there may be widespread agreement with a policy of supporting the increase in infrastructure investments for trade and economic growth, there are concerns in infrastructure construction. As reviewed earlier, even increasing investments in infrastructure would not bring about the expected results. In recent years, governments, especially those in developing countries, such as China and India, have initiated ambitious infrastructure investment plans (Arnold 2011). Empirical evidence in developing countries has shown that even if sufficient investments are raised, there are still a number of issues (for example delay, cost overruns, quality, safety and productivity) in infrastructure construction to transform these master plans into physical infrastructure capital (Le-Hoai et al. 2008; Toor and Ogunlana 2008; LaFraniere 2011; Memon et al. 2011). The poorly executed infrastructure investments require these issues in the implementation of infrastructure development plans to be examined.

5.4.2.1 Political Commitment

Studies on infrastructure development in East Asia have emphasized that sustained and powerful government leadership is crucial (Mody 1997). Infrastructure development involves a long-range vision that sustained commitment from the government is essential to support the development of a concrete strategy and subsequent actions. As a result, lack of political commitment could have direct effects on the implementation of infrastructure development plans. Literature on development planning highlights that the wide gap between plan formulation and plan implementation in most developing countries could be attributable to a lack of political will on the part of top leaders and high-level decision-makers (Waterston 2006). Lack of political commitment could also be the result of political discontinuity. Since the execution of national infrastructure projects often take a longer time than the length of one government, changes in the political affiliations of the government responsible for the projects create significant changes to the projects. Political inconsistencies at the national level and between different tiers of governments could also have adverse effects on the implementation of the projects. The lack of political consensus goes together with the lack of a high-powered government institution that provides an effective mechanism for implementing national infrastructure plans (Priemus 2010).

5.4.2.2 Problems of Corruption in Infrastructure Construction

Construction, in particular infrastructure construction, continues to be ranked as one of the most corrupt sectors worldwide. Corruption in the sector occurs in all stages from securing government contracts to the delivery of infrastructure. For securing government contracts, bribes are made to government officials responsible for budgeting decisions, project selections, tender specifications, procurement outcomes or contract negotiations and renegotiations. In the construction of infrastructure, bribes are made to government officials to cover poor-quality construction practices and outcomes, as well as theft of materials. Major impacts of corruption in infrastructure can lead to poor construction, limited occupational safety and low returns to government infrastructure investments (Kenny 2007). There are a number of causes of corruption in infrastructure construction, including the lack of transparency and competitiveness in bid processes, the discretionary power of individual bureaucrats involved in the award of contracts, inadequate financial and physical auditing, and inadequate capacity of regulatory bodies to enforce regulations (Kenny 2007; Dabla-Norris et al. 2011).

5.4.2.3 Land Acquisition

Problems in land acquisition can cause substantial delays and cost overruns in infrastructure construction (Priemus 2010). In many developing countries, land acquisition is considered one of the major barriers to governments' plans to develop infrastructure (Agrawal 1999; Morris 2007). Major problems in land acquisition for infrastructure in developing countries can include poor compensation and undervalued market price of land. Several causes for these problems can be identified, including lack of a negotiating mechanism to make land acquisition compensation more market oriented; bureaucracy in settling land disputes and claims; lack of a land acquisition compensation monitoring system; and lack of clarity about compensation valuation methods. Another problem in land acquisition is land price speculation. Increased costs of land acquisition caused by land price speculation could make infrastructure projects, especially toll road projects financially unfeasible. Lack of law enforcement to regulate the price of land acquired for infrastructure projects is regarded as the major cause of this problem (Chan 2003; Raghuram et al. 2009; Widhiarto 2011).

5.4.2.4 Building Capacity of Local Firms

As reviewed earlier in this study, various construction components, including finance, technology, management, materials and labour, are required in the construction of infrastructure projects. The level of development and the rate of implementation of infrastructure projects are determined by the combination and mobilization of these resources (Raftery et al. 1998). The inadequate capacity of the domestic construction firms in developing countries to meet the level of construction activities required for the construction of infrastructure could thus affect the implementation of infrastructure development plans. Moreover, the inadequate capacity of the domestic construction firms could lead to increasing foreign participation, which in turn could limit the opportunities for local firms to win contracts and for the local labour to gain employment. Inadequate capacity of local construction firms could therefore hurt other long-term development objectives, including employment generation and poverty alleviation. Improving capabilities and resources of local construction firms is thus necessary for achieving these development goals in developing countries (Wells and Hawkins 2010). Current issues pertaining to the capacity that domestic construction firms in developing countries are facing include poor level of efficiency and quality of work; poor level of professionalism and entrepreneurship; and resources shortages, especially in construction technology, management and finance (Ofori 2012; Hwang and Ng 2013). These problems become acute during periods of high economic activities, when demand for construction outputs, especially large infrastructure facilities, is higher (Howes and Robinson 2005).

5.4.2.5 Institutional and Legal Weaknesses in Infrastructure Construction

Other concerns involve institutional and legal weaknesses in infrastructure construction, including obsolescence of building regulations, changing and inconsistent law and regulations, ineffectiveness in the implementation of existing statutes and codes, and bureaucracy in formal procedures relating to project planning, construction permissions and administration (Raftery et al. 1998; Ofori 2000, 2006).

5.5 Conclusions

As reviewed earlier in this study, there is much literature on the relationship between investments in infrastructure and trade sector, as well as the national economy as a whole. To meet the huge demand for infrastructure, increasing infrastructure investment is encouraged in developing countries. The objective of investments in trade-related infrastructure is to stimulate trade, and thus economic growth. Besides the provision of infrastructure, investments in infrastructure construction are also used by governments in many developing countries as a tool for employment generation, increased work for local firms, and subsequently poverty alleviation and economic growth. As a result, public investments in infrastructure in developing countries have been increasing. However, empirical evidence has shown that these development objectives might not be realized as expected due to the low efficiency and quality of administering these public investments. Several studies have found that the low efficiency and quality of public investments in

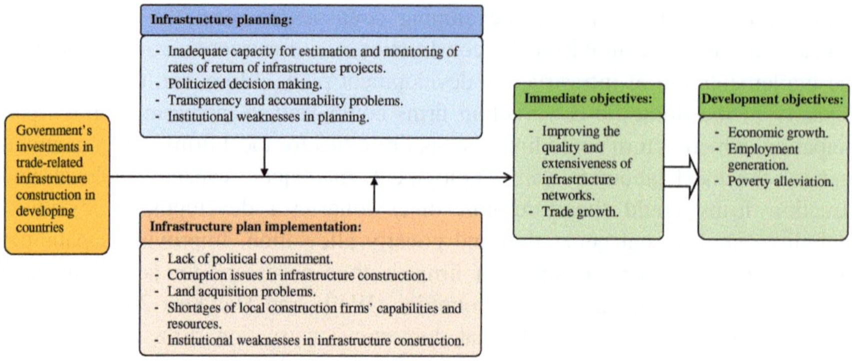

Fig. 5.1 Factors affecting the efficient use of infrastructure investments for trade and economic development from the literature review

infrastructure could be attributed to deficiencies in the planning and policy-making process (Henisz and Zelner 2006; Flyvbjerg 2007; Guasch et al. 2007; Collier and Venables 2008; Priemus 2010).

In addition, there are concerns about the implementation of infrastructure development plans that could affect the quality and efficiency of public investments in infrastructure (Kenny 2007; Dabla-Norris et al. 2011). This highlights the importance of bringing these issues in both planning and implementation processes together and looking at them through the role of the government as coordinator and facilitator of overall economic development. A number of issues were identified, notably inadequate capacity for estimating and monitoring of rates of return of infrastructure projects; politicized decision-making; transparency and accountability problems; institutional weaknesses in decision-making; lack of political commitment in the implementation of infrastructure development plans; corruption issues in infrastructure construction; problems in land acquisition; shortages of local construction firms' capabilities and resources; and institutional and legal weaknesses in infrastructure construction (see Fig. 5.1). In order to improve the efficiency of government funding invested in infrastructure to support trade and economic development, these issues must be resolved.

References

Agrawal, P. (1999). Urban land consolidation: A review of policy and procedures in Indonesia and other Asian countries. *GeoJournal, 49*(3), 311–322.

APEC Economic Committee. (1997). *Infrastructure development in APEC*. Paper presented at the 1997 Public-Business/Private Sector Dialogue, Los Cabos, Mexico.

Arnold, M. (2011). Infrastructure overview: The backbone of the global economy. Retrieved July 12, 2011 http://campaign.etfsecurities.com/aem/clients/ETF001/images/Infrastructure March2011.pdf.

Brooks, D. H. (2008). *Infrastructure and trade in Asia*. Cheltenham, UK: Edward Elgar.

Chan, N. (2003). Land acquisition compensation in China-problems and answers. *International Real Estate Review, 6*(1), 136–152.

Collier, P., & Venables, A. J. (2008). *Managing resource revenues: Lessons for low income countries*. Paper presented at the African Economic Research Consortium 2008 annual conference.

Dabla-Norris, E., Brumby, J., Kyobe, A., Mills, Z., & Papageorgiou, C. (2011). *Investing in public investment: An index of public investment efficiency* (Working Paper). Washington, D.C.: International Monetary Fund.

Davis, H. A. (Ed.). (2008). *Infrastructure finance: Trends and techniques*. London: Euromoney Books.

Devarajan, S., & Swaroop, V. (1993). *What do governments buy? The composition of public spending and economic performance* (Policy Research Working Paper No. 1082). Washington, D.C.: World Bank.

Devarajan, S., Swaroop, V., & Zou, H. (1996). The composition of public expenditure and economic growth. *Journal of Monetary Economics, 37*(2), 313–344.

Economic Research Center. (2001). *Transport and economic development*. Paper presented at the European Conference of Ministers of Transport—Report of the 119th Round Table on Transport Economics, Paris, France.

Estache, A. (2006). PPI partnerships vs. PPI divorces in LDCs. *Review of Industrial Organization, 29*(1), 3–26.

Estache, A., & Fay, M. (2007). *Current debates on infrastructure policy* (Policy Research Working Paper No. 4410). Washington, D.C.: World Bank.

Fay, M., & Morrison, M. (2008). Infrastructure needs in Latin America and the Caribbean. In H. A. Davis (Ed.), *Infrastructure finance: Trends and techniques* (pp. 48–75). London: Euromoney Books.

Fay, M., & Yepes, T. (2003). *Investing in infrastructure: What is needed from 2000 to 2010?* (World Bank Policy Research Working Paper No. 3102). Washington, D.C.: World Bank.

Flanagan, R., & Jewell, C. A. (2009). *Economics, urbanisation and infrastructure*. Geneva: FIDIC —International Federation of Consulting Engineers.

Flyvbjerg, B. (2007). Megaproject policy and planning: problems causes, cures. *Environment and Planning B: Planning and Design, 34*, 578–597.

Flyvbjerg, B., Holm, M. S., & Buhl, S. (2002). Underestimating costs in public works projects: Error or lie? *Journal of the American Planning Association, 68*(3), 279–295.

Glăvan, B. (2008). Coordination economics, poverty traps, and the market process: A new case for industrial policy? *The Independent Review, 13*(2), 225–243.

Guasch, J. L., Laffont, J. J., & Straub, S. (2007). Concessions of infrastructure in Latin America: Government-led renegotiation. *Journal of Applied Econometrics, 22*(7), 1267–1294.

Henisz, W. J., & Zelner, B. A. (2006). Interest groups, veto points, and electricity infrastructure deployment. *International Organization, 60*(1), 263–286.

Hoff, K. (2000). Beyond Rosenstein-Rodan: The modern theory of coordination problems in development. In B. Pleskovic & N. Stern (Eds.), *Annual World Bank Conference on Development Economics* (pp. 145–188). Washington, D.C.: World Bank.

Hossain, L. (2000). Infrastructure development planning in SEA: An analysis of the national telecommunications planning practice in Thailand. *Technovation, 20*(5), 275–285.

Howes, R., & Robinson, H. (2005). *Infrastructure for the built environment: Global procurement strategies*. Oxford: Butterworth-Heinemann.

Hulten, C. R. (1996). *Infrastructure capital and economic growth: How well you use it may be more important than how much you have* (NBER Working Paper No. 5847). Cambridge: National Bureau of Economic Research.

Hwang, B. G., & Ng, W. (2013). *Are Project Managers Ready for Green Construction?—challenges, knowledge areas, and skills*. Paper presented at the CIB World Building Congress 2013, Queensland, Australia.

Hwang, B.-G., Zhao, X., & Gay, M. J. S. (2012). Public private partnership projects in Singapore: Factors, critical risks and preferred risk allocation from the perspective of contractors. *International Journal of Project Management, 31*(3), 424–433.

Islam, N. (1990). *Horticultural exports of developing countries: Past performances, future prospects, and policy issues* (Research Report No. 80). Washington, DC: International Food Policy Research Institute.

Kenny, C. (2007). *Construction, corruption, and developing countries* (Policy Research Working Paper No. 4271). Washington, D.C.: World Bank.

Kocherlakota, N. R., & Yi, K. M. (1996). A simple time series test of endogenous vs. exogenous growth models: An application to the United States. *The Review of Economics and Statistics, 78*(1), 126–134.

Kohsaka, A. (2006). Overview: Infrastructure development in the Pacific region. In A. Kohsaka (Ed.), *Infrastructure development in the Pacific region* (pp. 1–20). London: Taylor & Francis.

Krugman, P. R. (1995). *Development, geography, and economic theory*. Cambridge: MIT Press.

Kumar, N. (2001). *Infrastructure availability, foreign direct investment inflows and their export-orientation: A cross-country exploration*. Paper presented at the Third Global Development Network Conference. Retrieved from http://www.gdnet.org/CMS/conference/papers/3rd_day2_6_nkumar.pdf.

LaFraniere, S. (2011, July 28, 2011). Chinese officials blame design flaws in deadly train crash. *The New York Times*. Retrieved from http://www.nytimes.com/2011/07/28/world/asia/28trains.html.

Le-Hoai, L., Lee, Y. D., & Lee, J. Y. (2008). Delay and cost overruns in Vietnam large construction projects: A comparison with other selected countries. *KSCE Journal of Civil Engineering, 12*(6), 367–377.

Low, S. P., & Dang, T. H. G. (2012). Role of marketing and construction in economic development: Lessons for emerging economies. *Business Review, 7*(1), 6–27.

Marshall, T. (2011). Reforming the process for infrastructure planning in the UK/England 1990–2010. *Town Planning Review, 82*(4), 441–467.

Matsuyama, K. (1997). Economic development as coordination problems. In M. Aoki, H. K. Kim, & M. Okuno-Fujiwara (Eds.), *The role of government in East Asian economic development: Comparative institutional analysis* (pp. 134–160). New York: Oxford University Press.

Memon, A. H., Abdul Rahman, I., Abdullah, M. R., Azis, A., & Asmi, A. (2011). *Assessing the effects of construction delays on MARA large projects*. Paper presented at the International conference on advanced science, engineering and information technology. Retrieved from http://eprints.uthm.edu.my/1662/.

Mody, A. (1997). Infrastructure delivery through central guidance. In A. Mody (Ed.), *Infrastructure strategies in East Asia: The untold story* (pp. xi–xxvii). Washington, D.C.: World Bank Publications.

Morris, S. (2007). Towards reform of the land acquisition framework in India. Indian Institute of Management working paper (2007–05), 04.

Morrissey, O., & Rudaheranwa, N. (1998). *Ugandan trade policy and export performance in the 1990s* (CREDIT discussion paper No. 6). Nottingham: University of Nottingham, Centre for Research in Economic Development and International Trade.

Mulu, T., & Smith, N. J. (2008). *Innovative road infrastructure financing in Jamaica and the developing countries*. Paper presented at the ARCOM 2008—the 24th annual conference, Cadiff, UK.

Murphy, K., Shleifer, A., & Vishny, R. (1989). Industrialization and the big push. *Journal of Political Economy, 97*(5), 1003–1026.

Mustajab, M. (2009). *Infrastructure investment in Indonesia: Process and impact. Dissertation*, University of Groningen, Nederlands. Retrieved from http://irs.ub.rug.nl/ppn/318943859.

Nwankwo, S. (2000). Assessing the marketing environment in sub-Saharan Africa: Opportunities and threats analysis. *Marketing Intelligence & Planning, 18*(3), 144–153.

Ofori, G. (2000). *Challenges of construction industries in developing countries: Lessons from various countries*. Paper presented at the 2nd international conference on construction in developing countries. Retrieved from http://rec.hku.hk/steve/Msc/08-09/reco6005/RFF_challenges.pdf.

Ofori, G. (2006). Revaluing construction in developing countries: A research agenda. *Journal of Construction in Developing Countries, 11*(1), 1–16.

Ofori, G. (Ed.). (2012). *New perspectives on construction in developing countries*. UK: Spon Press.

Ohlin, B. G. (1933). *Interregional and international trade*. Cambridge: Harvard University Press.

Peters, H. (1992). *Service: The new focus in international manufacturing and trade* (Policy Research Working Paper Series No. 950). Washington, D.C.: The World Bank.

Porter, M. E. (1990). *The competitive advantage of nations*. New York: Free Press.

Porter, M. E. (1998). Clusters and the new economics of competition. *Harvard Business Review, 76*(6), 77–90.

Porter, M. E. (2000). Location, competition, and economic development: Local clusters in a global economy. *Economic development quarterly, 14*(1), 15–35.

Priemus, H. (2010). Mega-projects: Dealing with Pitfalls. *European Planning Studies, 18*(7), 1023–1039.

Raftery, J., Pasadilla, B., Chiang, Y., Hui, E., & Tang, B. (1998). Globalization and construction industry development: Implications of recent developments in the construction sector in Asia. *Construction Management and Economics, 16*(6), 729–737.

Raghuram, G., Bastian, S., & Sundaram, S. S. (2009). *Mega projects in India environmental and land acquisition issues in the road sector* (Working Paper No. 2009-03-07). India: Indian Institute of Management Ahmedabab.

Ricardo, D. (1971). *On the principles of political economy, and taxation*. Harmondsworth: Penguin.

Rosenstein-Rodan, P. N. (1943). Problems of industrialisation of eastern and south-eastern Europe. *The Economic Journal, 53*(210, 211), 202–211.

Schweikert, A., & Chinowsky, P. (2012). *National infrastructure planning: A holistic approach to policy development in developing countries*. Paper presented at the Engineering Project Organizations conference, Rheden, The Netherlands.

Short, J., & Kopp, A. (2005). Transport infrastructure: Investment and planning. Policy and research aspects. *Transport policy, 12*(4), 360–367.

Smith, A. (1976). *An inquiry into the nature and causes of the wealth of nations*. Oxford: Clarendon Press.

Smith, N. J., & Gannon, M. (2008). Political risk in light rail transit projects. *ICE Proceedings in Management Procurement and Law, 161*(4), 179–185.

Taylor, A. M. (2002). *Globalization, trade, and development: Some lessons from history* (NBER Working Paper No. w9326). Cambridge, MA: National Bureau of Economic Research.

Thirlwall, A. P. (2000). *Trade, trade liberalisation and economic growth: Theory and evidence* (Economic Research Paper No. 63). Côte d'Ivoire: African Development Bank.

Todaro, M., & Smith, S. (2003). *Economic development* (8th ed.). Boston: Addison Wesley.

Todaro, M., & Smith, S. (2009). *Economic development* (10th ed.). Boston: Addison Wesley.

Toor, S. U. R., & Ogunlana, S. (2008). Problems causing delays in major construction projects in Thailand. *Construction Management and Economics, 26*(4), 395–408.

Venables, A. (2004). Infrastructure, trade costs and gains from international trade. In European Conference of Ministers of Transport (Ed.), *Report of the 130th Round Table on Transport Economics: Transport and International Trade* (pp. 57–82). Paris, France: European Conference of Ministers of Transport.

Waterston, A. (2006). Development planning: Lessons of experience. In E. E. Otenyo & N. S. Lind (Eds.), *Comparative public administration (research in public policy analysis and management* (Vol. 15, pp. 427–431). Emerald Group Publishing Limited.

Wells, J., & Hawkins, J. (2010). Increasing 'local content' in infrastructure procurement. Part I. *Proceedings of the Institution of Civil Engineers Management, Procurement and Law, 163* (MP2), 65–70.

Widhiarto, H. (2011, April 25). Land acquisition remains a tough nut to crack. *The Jakarta post*. Retrieved from http://www.thejakartapost.com/news/2011/04/25/land-acquisition-remains-a-tough-nut-crack.html.

World Economic Forum. (2007). *Building centre of excellence*. Paper presented at the 23rd India Economic Summit, New Delhi.

World Economic Forum. (2012). *Strategic infrastructure—steps to prioritize and deliver infrastructure effectively and efficiently*. Retrieved from http://www.weforum.org/reports/strategic-infrastructure-steps-prioritize-and-deliver-infrastructure-effectively-and-efficie.

Chapter 6
Vietnam—Economic, Trade and Infrastructure Overview

Abstract This chapter provides an overview of economic development in Vietnam over the past decades. Emphasis is placed on the roles of trade and infrastructure construction in economic development, and how well the development of these sectors is complementary to each other. In addition, the chapter provides an understanding of Vietnam's planning process at the national level, and major tasks of government agencies are related to trade, infrastructure planning and development.

Keywords Vietnam · Economic development · Infrastructure development · Trade · Government planning

6.1 Introduction

Since Vietnam started its economic reforms in 1986, the country has gone from one of the poorest in the world to become a lower-middle-income country (World Bank Data 2013). Moving forward, the country faces new challenges in economic development. The chapter provides an overview of economic development in Vietnam over the past decades. Emphasis is placed on the roles of trade and infrastructure construction in economic development, and how well the development of these sectors is complementary to each other. In addition, the chapter provides an understanding of Vietnam's planning process at the national level, and major tasks of government agencies are related to trade, infrastructure planning and development.

6.2 Vietnam—Country and Economic Overview

6.2.1 Geography and Population

Vietnam is located in Southeast Asia, bordering China in the north, Cambodia and Laos in the west, and the South China Sea and Pacific Ocean in the east and south (see Fig. 6.1 adapted from the Government Web Portal of Vietnam, 2014).

G. Dang and L. Sui Pheng, *Infrastructure Investments in Developing Economies*,
DOI 10.1007/978-981-287-248-7_6

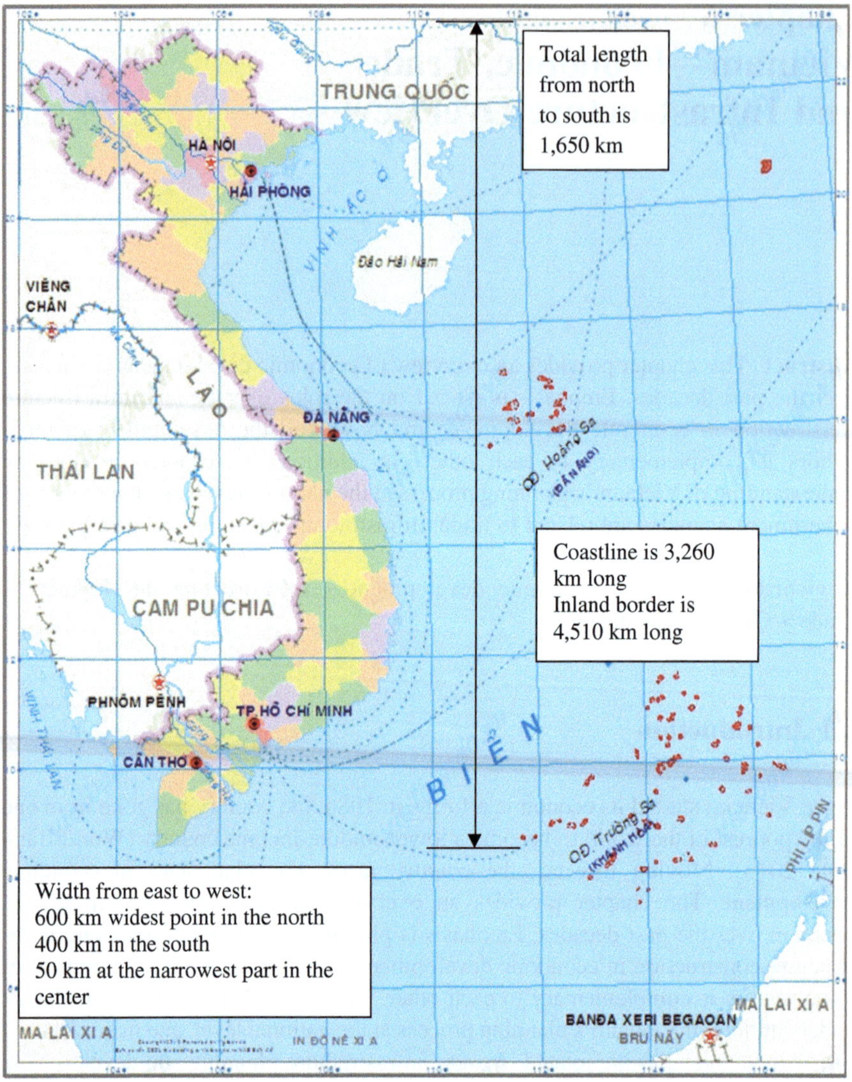

Fig. 6.1 Geographical map of Vietnam

The country has a total land area of more than 331,000 km^2 and a coastline of more than 3,000 km lying on international seaways running along the country from the broad plain of the Red River and the Gulf of Tonkin in the north through to the Mekong Delta in the south (Government Web Portal of Vietnam 2010a). In 2009, total population was estimated as 86 million people. About 24 % of the population is under the age of 15, and 69 % is from the age of 15–64. Population growth rate averaged at about 1.2 % per year from 1999 to 2009. Adult literacy rate is 94 % (United Nations 2010a).

The urban population increased rapidly with an average annual growth rate of about 3.4 %. At present, urban population accounts for about 29.6 % of the total population. Vietnam is divided into 63 cities and provinces. The two largest socio-economic centres are Hanoi, the capital city in the north with a population of about 6.5 million people, and Ho Chi Minh City in the south with a population of about 7.16 million (data compiled from the General Statistics Office of Vietnam, GSO 2010). Other key urban areas include the port cities of Hai Phong in the north, Da Nang on the Central Coast and Can Tho in the Mekong Delta in the south (Government Web Portal of Vietnam 2010a).

6.2.2 Economic Overview

Defeating the French and Japanese colonial forces, the Democratic Republic of Vietnam was established in 1945. However, it was not until 1975 that the north and the south of Vietnam were reunified. In 1976, the country was renamed the Socialist Republic of Vietnam (Government Web Portal of Vietnam 2010b). Since then, the country started its reconstruction and has made remarkable achievements. The country has recently become a lower-middle-income country (World Bank Data 2013). The growth record has been largely the results of the country's economic reforms initiated in the 1980s.

After independence, Vietnam faced tremendous tasks to restore the country's economy that was damaged by the war. The first five-year socio-economic development plan was formulated for the period 1960–1965 for the north. The second five-year (1976–1980) plan was prepared after the reunification in 1975; thus, this is considered the first national five-year plan. However, both plans pursued socialist industrialization with a major focus on heavy industry and rapid agricultural growth with the former placed at the forefront of development (Communist Party of Vietnam 2005). Between 1976 and 1980, the economy grew by only 0.4 % per year, industrial production by 0.6 % and agricultural outputs by 1.9 %, and living conditions remained poor. The next five-year (1981–1985) plan continued to place its focus on heavy industry and agricultural development, although the development of agriculture was shifted to the forefront (Vo et al. 1995; Communist Party of Vietnam 2005). Nonetheless, the results were limited with inflation soaring to 700 % in 1986 and the then socio-economic crisis was becoming serious at the end of this period (Harvie and Tran 1997). The failure was considered to have resulted primarily from the centrally planned, subsidy-based and relatively autarkic framework that Vietnam was pursuing (Vo et al. 1995). Realizing these problems, the Sixth National Party Congress in December 1986 approved a radical reform, "Doi Moi", i.e. "Renovation", thus initiating Vietnam's gradual transition from a centrally planned economy to a socialist-oriented market economy. Major policy changes initiated by Doi Moi were as follows:

- The decentralization of state economic management to enterprise level and the provision of autonomy for state-owned enterprises (SOEs) in making decisions relating to production, distribution and financing.
- The replacement of administrative measures and controls with economic tools to deal with macroeconomic issues, particularly the use of market-oriented monetary policies to control inflation.
- The adoption of an outward-oriented policy in external economic relations, key elements of this policy including currency devaluation and the introduction of foreign investment law.
- The recognition of the private sector as the engine of economic growth, key elements of this policy included decollectivization of agriculture and legal recognition of private business (Than and Tan 1993).

It is noticeable that the central theme of Doi Moi is market liberalization. Vo Nguyen Giap, the then Vice-Chairman of the Council of Ministers of the Socialist Republic of Vietnam, regarded Doi Moi as the one blueprint that has marked a new development in Vietnam's thinking on economic theory and in the country's socio-economic policies (as cited by Ronnås and Sjöberg 1990). The results of these efforts started to be realized at the end of the next following five-year plan (1986–1990). The central economic objectives of the 1986–1990 plan were to increase the production of food, consumer goods and export goods. The most recognized achievement of the plan was the blooming of agriculture production. Since 1989, agriculture production not only put an end to the country's persistent shortage of food, but also transformed Vietnam from a net food importer to a major world rice exporter. Inflation rate also fell from 300 % in 1987 to 67.1 % in 1991. The economy was able to grow at an average growth rate of 3.7 % per year between 1987 and 1991, although the socio-economic crisis had not yet ended (Vo et al. 1995; Chiang et al. 2004).

Following the social and economic policy reforms of Doi Moi, the next phase in the renovation course in economic development of Vietnam was set up by the first socio-economic development strategy (SEDS) for 1991–2000. 1991–2000 SEDS was named "Socio-Economic Stabilization and Development Strategy to 2000" (Ministry of Planning and Investment 2007). The overall objectives set in the 1991–2000 SEDS were:

- to overcome economic crisis and to stabilize the socio-economic situation;
- to rise above poverty and underdevelopment and to improve living standards;
- to double the country's GDP in 1990 by 2000;
- to establish an open economic system;
- to create a multi-sectoral economy;
- to enhance national defence and security; and
- to create the necessary conditions for the country to develop more rapidly in the early twenty-first century (Communist Party of Vietnam 2005).

Based on these strategic goals of the 1991–2000 SEDS, the two five-year socio-economic development plans, 1991–1995 and 1996–2000, were prepared with more specific goals such as those to keep inflation under control; to boost production; to increase accumulation; to gradually improve living standards; to ensure macro-economic stabilization; to prepare for the next development phase with focus on human resources, science and technology, infrastructure and institutional framework; to develop multi-sector economy; to extend international cooperation; and to combine economic development with national defence and security (Communist Party of Vietnam 2005). These plans had not only driven Vietnam out of the socio-economic crisis, but also pushed the economy to grow rapidly since 1992. The economy grew at an average rate of about 7 % per annum between 1991–2000, nearly double that of the period 1986–1990 (data compiled from the United Nations Statistics Division, UNSD 2010). The highest growth during the period was 9.5 % in 1995. Hyperinflation was curbed. Inflation rate fell dramatically from 67.5 % in 1990 to 12.2 % in 1995 (Harvie and Tran 1997). Share of domestic savings in GDP reached 25 % in 2000. Living standards had improved with people living below the poverty line (internationally defined) falling from 70 % in the mid-1980s to 37 % in 2000, according to the United Nations Industrial Development Organization (UNIDO) (as cited by Chiang et al. 2004).

Results of the first step towards industrialization had also been realized during the period. Share of agriculture production in GDP at constant (1994) prices fell from around 32 % in 1990 to around 23 % in 2000, and that of manufacturing and construction rose from 25 % in 1990 to around 35 %, while there had not been much change in that of the service sector. The manufacturing and construction sector was the fastest-growing sector with an average annual growth rate of 11.3 % between 1991 and 2000. The service sector came in second with 7.1 % and agriculture only 4.2 %. The expanding growth in this period was also supported by strong performance in exports and foreign investments. Exports of goods averaged an annual growth rate of around 20 %. Trade deficit over total export value fell from about 50 % in 1995 to 8 % in 2000 (data compiled from the General Statistics Office of Vietnam 2010). Net inflows of registered foreign direct investments (FDI) reached an average value of USD 3.7 billion per year (Vo 2005). A multi-sector economy had thus been gradually shaped with expansion in the domestic and foreign private sectors. During this period, international cooperation was also strengthened, especially when Vietnam joined the Association of Southeast Asian Nations (ASEAN) in 1995 and the Asia-Pacific Economic Cooperation (APEC) Forum in 1998. In general, most of the main targets of the 1991–2000 SEDS had been achieved (Communist Party of Vietnam 2005).

Vietnam was, however, still a poor and under-developed country at the end of the period 1991–2000. Gross National Income (GNI) per capita was only USD 390 in 2000 (World Bank 2010). The development of Vietnam still lagged far behind the average world standards and many neighbouring countries. Addressing these issues, the 2001–2010 SEDS outlined the following long-term development goals for Vietnam:

- to come out of underdevelopment;
- to improve substantially the living standards;

- to lay the foundation for a modern-oriented industrialized country by 2020;
- to enrich human resources, scientific and technological capacities;
- to build infrastructure for economic development as well as national defence and security;
- to establish necessary institutions for a socialist-oriented market economy; and
- to raise the country's status in the international community (Communist Party of Vietnam 2005).

Implementing the 2001–2010 SEDS (with two five-year plans, 2001–2005 and 2006–2010), Vietnam has been successfully moving forward with a good number of significant achievements. Vietnam has been able to maintain a relatively high economic growth of nearly 8 % per year from 2001 to 2007 before slowing down to 6.3 % in 2008 due to the 2008 global financial crisis (see Fig. 6.2) (United Nations Statistics Division 2010).

The economic structure continued to move towards the manufacturing and construction sector (see Fig. 6.3). Share of agricultural production in GDP further declined from 23 % in 2000 to 18 % in 2008, that of the manufacturing and construction sector increased from 35 % in 2000 to 45 % in 2008, and that of the service sector maintained a portion of around 41 %. The manufacturing and construction sector was still the fastest-growing sectors with an average annual growth rate of 9.7 % from 2001–2008, that of the service sector and agriculture were 7.4 and 3.9 %, respectively (General Statistics Office of Vietnam 2010).

Total value of exports of goods continued to grow at about 20 % per year from 2001 to 2008. The value in 2008 was USD 62.7 billion, which has more than quadrupled that of 2000. Net inflows of FDI increased steadily between USD 1.3 billion and USD 1.8 billion from 2002 to 2006, but soared to USD 6.6 billion in

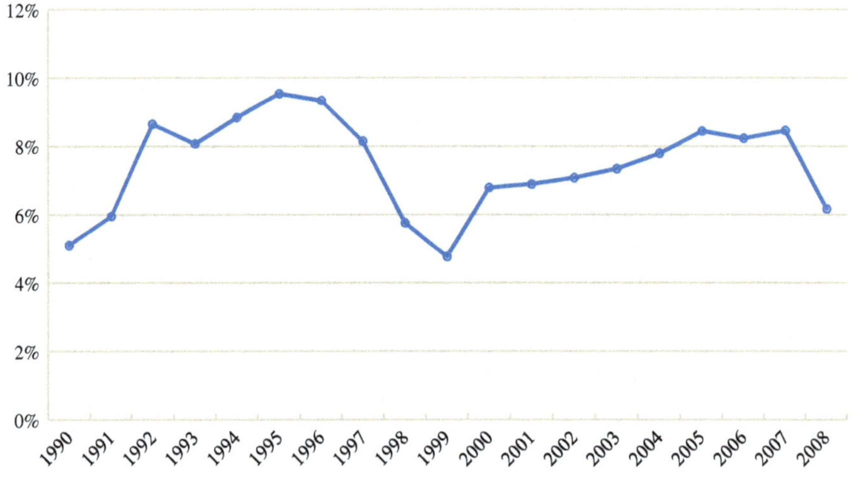

Fig. 6.2 Vietnam's GDP growth (at constant 1990 prices US$). *Source* Data compiled from the United Nations Statistics Division (2010)

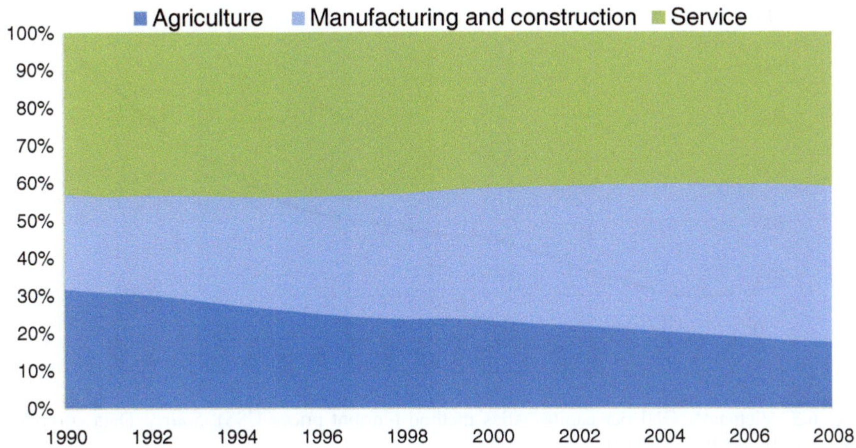

Fig. 6.3 Structure of Vietnam's GDP at 1994 prices by economic activities. *Source* Data compiled from the General Statistics Office of Vietnam (2010)

2007 and USD 9.3 billion in 2008 (Asian Development Bank 2009). In addition, a young and vibrant domestic private sector has been growing (see Fig. 6.4). The sector accounts for 33 % of manufacturing output and about 90 % of labour market annually (Vo 2007).

The social aspects of development have been enhanced as well. The country has seen a dramatic increase in GNI per capita from USD 130 in 1990 to USD 1010 in 2009 (see Fig. 6.5), thus moving from one of the poorest countries in the world to a

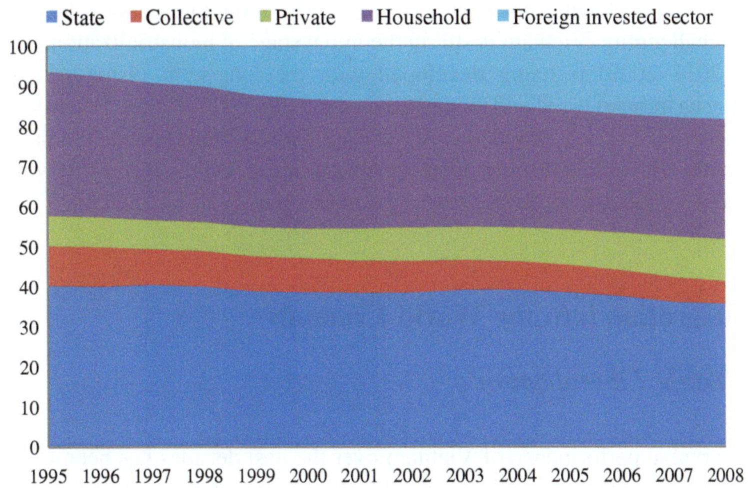

Fig. 6.4 Structure of Vietnam's GDP at current prices by ownership (%). *Source* Data compiled from the General Statistics Office of Vietnam (2010)

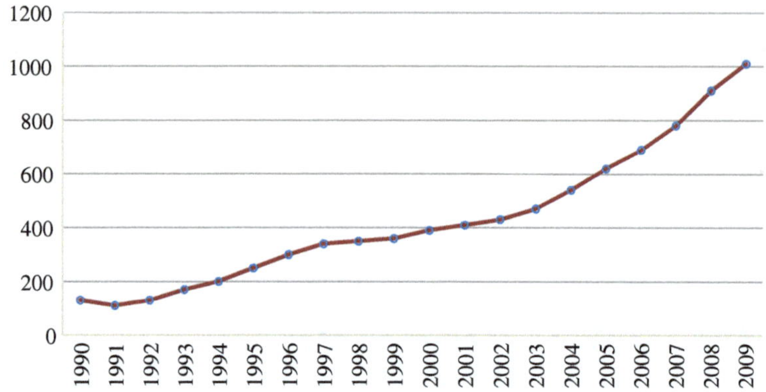

Fig. 6.5 Vietnam's GNI per capita, Atlas method (current prices US$). *Source* Data compiled from the World Bank (2010)

lower-middle-income country going by the World Bank classification method (World Bank 2010). Poverty has been reduced considerably with the population living below the poverty line decreasing from 37 % in 2000 to 12 % in 2009 (United Nations 2010b). International cooperation during this period also attained some major achievements when Vietnam signed the permanent normal trade relations (PNTR) with the USA in 2000 and became a member of the World Trade Organization (WTO) in 2007 (Kien 2010).

Vietnam has reached the end of its 2001–2010 SEDS. In general, the two previous SEDS (1991–2000 and 2001–2010) have driven Vietnam from a largely poor, centrally planned and agriculture-based economy to a lower-middle-income, rapidly growing, market-based economy and one which is increasingly being integrated into the regional and global community. However, the economic outlook for Vietnam remains challenging. Vietnam is still in the initial stage of industrialization. The new SEDS (2011–2020) is being developed with the long-term vision of Vietnam becoming an industrialized and modernized country by 2020 (United Nations 2010b). The next period 2011–2020 will therefore be particularly important for Vietnam to lay the groundwork for that turning point. Further progress will be increasingly difficult and need to be guided by proper strategies, action plans and institutions.

6.3 Integration into the World Economy

6.3.1 Trade Liberalization

The remarkable performance of Vietnam over the past decades has been much of a result from an outward-oriented approach for economic development that Vietnam has adopted since Doi Moi in 1986. Before Doi Moi, trade opportunities and specialization potentials were neglected in the first two five-year plans (1961–1965 and 1976–1980). These plans assigned priority to achieve self-sufficiency in heavy industries. The poor performance of these plans resulted in a modification in economic orientation of the 1981–1985 five-year plan. The focus of the 1981–1985 plan was placed on agriculture and light industries rather than heavy industries (Reinhardt 1993; Vo et al. 1995). Nevertheless, the economic orientation of the 1981–1985 plan was still largely similar to the previous ones, which was to pursue socialist industrialization. A major shift in policy orientation was initiated in December 1986 with decisions on the adoption of an outward-oriented approach. The centre of the next five-year plan (1986–1990) was to increase production of food, consumer goods and export goods. Since then, trade opportunities and specialization have become part of the priority list of economic policy of Vietnamese government. In the first SEDS (1991–2000) of Vietnam, the overall goals for the trade sector were:

- to reduce share of primary products and to increase share of processed products in the structure of exports; to develop goods that have high total export turnover and stable markets;
- to develop various services, particularly tourism, which can generate foreign exchange;
- to import goods that effectively stimulate production development and technology modernization, which in turn can improve quality and competiveness of domestic goods, and meet domestic demand;
- to diversify markets and to develop multilateral trading ties, to vigorously penetrate into new markets and to shape a favourable position in the new markets; and
- to expand the trade rights; to eliminate non-tariff barriers; to reduce government interventions in business enterprises; to improve the roles of macro-economic tools such as taxes, interests and exchange rates; and to improve the legal environment (Prime Minister's Office of Vietnam 2000b).

The missions of the trade sector were further enhanced in the 2001–2010 SEDS with the overall goals of the trade sector included to boost export growth, in part to further stimulate industrialization and modernization, to generate employment and to earn foreign currencies; to shift the export structure to high value-added goods, processed and manufactured goods, and hi-tech products, and to stimulate service export; to focus on importing equipment and materials for production, especially advanced technologies, to maintain a reasonable balance of trade and gradually balance exports and imports; to expand and diversified markets and business modes; and to successfully integrate into regional and world economies (Prime Minister's Office of Vietnam 2000b). In order to achieve these goals, the following missions were identified for the trade sector:

- to continue to pursue the policy that gives high priority to exports in order to stimulate GDP growth, to develop production, to generate employment and to obtain foreign currencies;

- to actively integrate into regional and world economic areas, on the basis of protecting national sovereignty and the socialist-orientation, with a master plan and roadmap suitable for Vietnam's economic development progress, and regulations of international organizations of which Vietnam is a member;
- to consider internalizing values, to shift the economic structure and to review the management mechanism; to improve the legal environment; and to increase efficiency and competitiveness of firms as well as the national economy as key to the expansion of the trade sector and international integration;
- to join domestic markets with foreign ones; to focus on domestic markets along with expanding and diversifying foreign markets; and
- to continue to pursue the policy that diversifies economic sectors participating in trading activities with the state-sector playing the leading role.

Regulated by the Ministry of Trade (now the Ministry of Industry and Trade, MoIT), Vietnam's legal and institutional framework for trading activities and related policies have been much improved since Doi Moi. These significant changes consist of the removal of restrictions on private-sector trade; the elimination of the monopoly position of SOEs in trading activities; the reduction of quantitative control; the introduction of foreign investment law; the removal of foreign exchange requirements for foreign-invested enterprises (FIEs); the establishment of export-processing zones; the implementation of import tariffs with the preferential rates (lower tariff rates) applied to exported/imported goods to/from countries that have signed trade agreements with Vietnam; the introduction of export promotion measures such as zero export duty, tax exemption, export credit and especially the duty drawback scheme. Legal reforms have also been highlighted with the introduction of new enterprise, competition and investment laws since 2000 (Than and Tan 1993; Vo 2005; Abbott et al. 2009). Further efforts in reducing trade barriers have been made when Vietnam joined ASEAN in 1995 and APEC in 1998. Vietnam has also engaged in major bilateral, multilateral trade agreements and free trade agreements such as the Vietnam-EU trade agreement in 1992, the ASEAN Free Trade Area in 2001, the Vietnam-US bilateral trade agreement in 2000, the ASEAN-China Free Trade Area in 2002 and the ASEAN-Japan Free Trade Area (2003) (Vo 2007; Abbott et al. 2009). Early 2007 marked the country becoming the 150th member of the WTO. Nearly 220 countries and territories now have trade relations with Vietnam (Dinh 2006). The major timeline of these significant changes is shown in Table 6.1.

6.3.2 International Trade and Economic Growth

As a result of these improvements, achievements of the trade sector in Vietnam have been impressive. Size of trade has increased rapidly since Vietnam introduced Doi Moi in 1986. The value of total exports and imports of goods and services (at constant 2000 US$) rose dramatically from only about USD 6.1 billion in 1989 to

Table 6.1 Significant changes in legal and institutional framework for trading activities and related policies of Vietnam since Doi Moi

Year	Changes
1986	Outward-oriented approach was adopted
1987	Restrictions on private-sector trade were removed
	Foreign investment law was introduced
	The official exchange rate was devalued
1988	Import tariffs were introduced
	Strict control over foreign exchange was enacted
	The Vietnamese dong was sharply devalued to improve the country's trade performance
1989	Import duties for industrial needs were removed
	State monopoly in foreign trade was eliminated
	The State Committee for Co-operation and Investment (SCCI), an inter-ministerial body, was created to promote, approve, facilitate and implement all foreign investments in the country
	A decree to protect property rights was introduced
1991	Law on import and export duties was approved with the application of preferential rates of tariff applied to exported/imported goods to/from countries that have signed trade agreements with Vietnam
	The "Vietnam Investment Forum" to attract foreign investments was jointly organized by SCCI, UNIDO, and the United Nations Development Program (UNDP)
	Law on export-processing zones (EPZs) was introduced
1992	Trade agreement with the European Union was signed
1994	The US's trade embargo against Vietnam was lifted
1995	Vietnam joined ASEAN
	WTO accession working party was established
1997	The "Vietnam Business Forum" was created to facilitate communication between the government and the business community, including foreign businesses. The forum was chaired by a Deputy Prime Minister and co-chaired by the World Bank Country Director, the IFC Country Manager and the Minister for Planning and Investment
	Reduced requirements on firms to enter foreign trade
	Commercial law was introduced
1998	Trade licences were abolished
	Foreign exchange was required to be deposited in one onshore account by all economic entities
	Vietnam joined APEC
1999	Japan and Vietnam officially exchanged most favoured nation (MFN) treatment for tariffs
2000	Vietnam Trade Promotion Agency under the Ministry of Trade was established. The agency is responsible for state management, coordination and implementation of trade and related-trade promotion activities
	FIEs were allowed to get foreign currency loans from domestic banks to pay off loans from offshore banks
	The Vietnam-US bilateral trade agreement was signed

(continued)

Table 6.1 (continued)

Year	Changes
2001	A trade policy roadmap for the period 2001–2005 was announced for the first time, making the export–import environment more transparent and predictable
	All domestic enterprises have been allowed to trade commodities or items freely, except those prohibited or under specialized management
	Most quantitative restrictions were removed
	Commitments under CEPT/AFTA framework began to be implemented
	Customs law was introduced
2002	FIES have been granted the right to export goods other than those they produce
	Vietnam participated in ASEAN-China Free Trade Area
	Commitments under US-BTA started to be undertaken
2003	Tariff-rate quotas were introduced and applied for certain agricultural products such as cotton, tobacco materials and salt
	Restriction on foreign exchange deposits for all economic entities was removed.
	Vietnam participated in the ASEAN-Japan free trade area
2004	EU-Vietnam bilateral agreement on WTO accession was signed
	Competition Law was introduced
2005	29 new or amended laws on commerce and trade were introduced
	Enterprise law was introduced
	Customs law was revised to meet the requirements of a modern customs system
	Vietnam agreed to implement a single window approach by 2012 for the establishment of the ASEAN single window
2006	Final bilateral agreements for WTO accession were reached
	CEPT/AFTA implementation was completed
	Investment law was introduced
	Law on intellectual property rights was introduced
	E-transaction law and the decree on E-commerce came into effect
	The first master plan on E-commerce was implemented for 2005–2010
2007	Vietnam became a member of WTO

Source Abbott et al. (2009), National Assembly of Vietnam (1997, 2001), Harvie and Tran (1997), Reinhardt (1993), Than and Tan (1993), Vietnam Business Forum (2010), Vo (2005, 2007)

USD 35 billion in 2000, and to USD 93 billion in 2008. While trade as a percentage of GDP was only 23 % in 1986, the ratio rose to 103 % in 1999 and to 172 % in 2008. During that time, exports grew relatively faster than imports (Abbott et al. 2009). Contributions of exports to GDP went up from 6.6 % in 1986 to 50 % in 1999 and to 78 % in 2008. Similarly, the import dependency ratio increased from 16.6 % in 1986 to 52.8 % in 1999 and 94.7 % in 2008 (see Fig. 6.6).

The trade pattern has also been changing with the export composition shifting away from primary products and resource-based manufacturing towards labour-intensive light manufacturing, and more recently towards component production/assembly within the high-tech industries (Asian Development Bank 2009). Share of

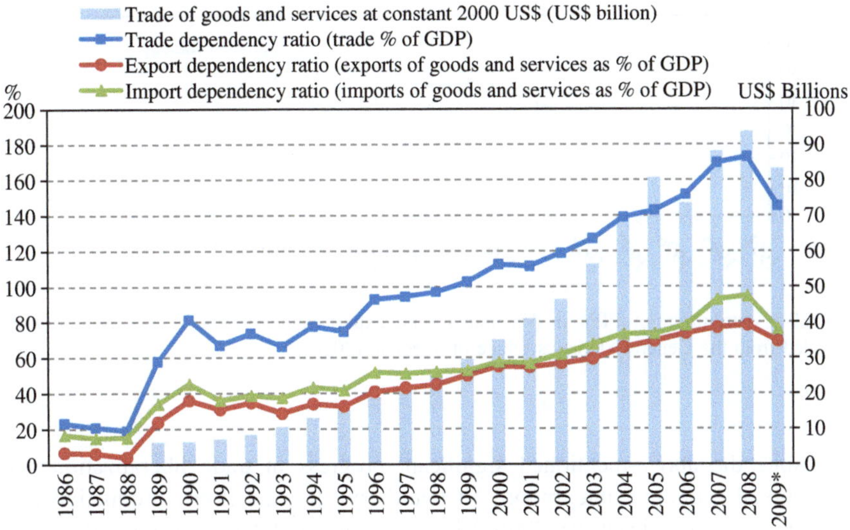

Fig. 6.6 Contribution of foreign trade to Vietnam's GDP. *Source* Data compiled from the World Bank (2010)

agriculture in total exports declined from 32 % in 1995 to 14.5 % in 2007, while that of light industry and handicrafts improved from 28 to 42 %, and heavy industries and minerals from 25 to 34 %. The trade markets have also been much more diversified since 1986. Before 1986, trade was concentrated in countries of the Soviet bloc. However, the trade markets have rapidly expanded to many other countries, especially in Asia, Europe and North America (Athukorala 2009; Das and Shrestha 2009). Major trading partners of Vietnam include ASEAN economies, Japan and China in Asia; Germany, United Kingdom and France in Europe; the USA in North America; and Australia in Oceania. Exports to G3 economies, Japan, Europe and USA, account for nearly 50 % of total export value (data compiled from the General Statistics Office of Vietnam 2010).

During the same period of time, the country has been able to attract massive inflows of FDI, of which a significant portion was export-oriented (Athukorala 2009). The share of FDI in total investments improved significantly in the mid-1990s and reached 30 % in 1995. Affected by the 1997–1998 Asian financial crisis, the share went down and remained stable around 17–18 % at the end of the 1990s and the first half of the 2000s or equivalently an increase by USD 1.3–1.8 billion in net inflows of FDI per year. However, the value soared to USD 6.6 billion in 2007 and USD 9.3 billion in 2008. Accordingly, the share of FDI in total investments increased to 27 % in 2007 and 35 % in 2008. The contribution of FIEs in GDP expanded from only 2 % in 1992 to 13 % in 2000 and to 18 % in 2008 (Vo 2007; Asian Development Bank 2009; Athukorala 2009; Vo and Nguyen 2009; General Statistics Office of Vietnam 2010). Share of FIEs in total manufacturing exports increased from about 20 % in 1991 to 47 % in 2000 and to 55 % in 2008 (see Fig. 6.7).

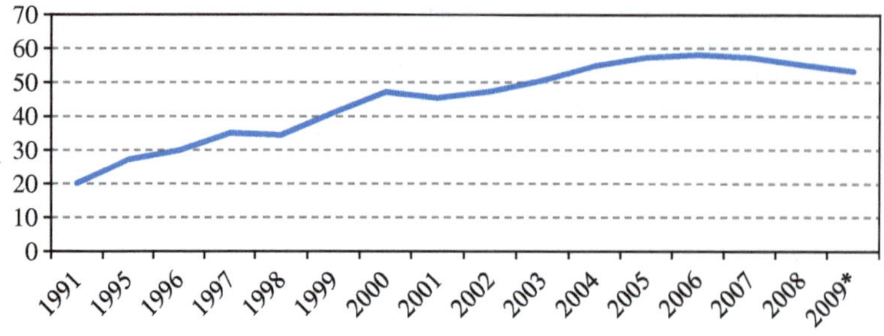

Fig. 6.7 Contribution of FIEs in Vietnam's total export value of goods (%). *Source* Data compiled from the General Statistics Office of Vietnam (2010)

Despite the above-mentioned significant progress in trade liberalization and international economic integration, Vietnam's position in global production networks is still insignificant. By 2006, Vietnam's share in total world non-oil exports and in total non-oil exports of developing countries were only 0.34 and 0.9 %, respectively. Compared to the other five major ASEAN member countries (Indonesia, Malaysia, Philippines, Singapore and Thailand), Vietnam has the smallest share in total world non-oil exports. Vietnam's 2006 export level of about USD 41 billion was surpassed by regional countries such as Taiwan, South Korea, Malaysia, Indonesia, Thailand and the Philippines in the 1980s and the 1990s (Athukorala 2009). Moreover, the 2006 share of manufacturing (39 %) and hi-tech exports (4.5 %) in total exports of Vietnam, albeit growing, was relatively smaller than those of the East Asian countries between 1997 and 1998. Producing low value-added products, most private firms in Vietnam were small-sized and remained at very weak positions in the global and regional value chains (Vo 2007).

There have been some concerns over the effectiveness of FDI recently. Large inflows of FDI were considered a necessary condition for Vietnam's economic transition (Ohno 2009). However, in the past few years, the gap between registered and realized FDI has increased. The disbursement rates of FDI were 41 % in 2006, 31 % in 2007 and 18 % in 2008, much lower than the average of 52.7 % during the 1988–2007 period. Moreover, FIEs in heavy industries such as chemicals and chemical products, basic metal products, fabricated metal products and motor vehicles have not contributed as much to export expansion as expected (Athukorala 2009). FDI's positive spillover effects via skills and technology transfer have not been fully realized. Vietnam's internal value, a sufficient condition for Vietnam in the next stage of development, remains small. Foreigners still dominate in all areas of production including management, technology, design, factory operations, logistics, quality control and marketing (Ohno 2009).

The obstacles to the above-mentioned constraints in the development of trade and international economic integration of Vietnam have been highlighted in a number of studies. The constraints are related much to the country's capacity bottlenecks in the domestic private sector, industrial human resources, institutions and infrastructure (Ohno 2009; Tran 2009; Vo and Nguyen 2009). Issues related to infrastructure in Vietnam are considered in the following section.

6.4 Infrastructure Development

6.4.1 Infrastructure-Related Indicators Compared to Other Asian Countries

Since a proper infrastructure system is recognized as crucial for Vietnam to sustain the high economic growth, infrastructure has been included in the national development plans. In the first SEDS (1991–2000), the strategy focused on upgrading and modernizing major communication and transport (including sea, land and air transport) networks that facilitate the North–South link and international links. In the second SEDS (2001–2010), the first priority was to accelerate the development of the energy sector. In order to realize these tasks, the Vietnamese government has invested billions of dollars in infrastructure. Over the last decade, Vietnam has sustained about 9–10 % of GDP invested in infrastructure (Nguyen and Dapice 2009; Moore et al. 2010). Compared to other East Asian countries during their period of rapid industrialization, Vietnam's level of infrastructure investment was relatively higher. For example, Taiwan invested 9.5 % of GDP during 1970–1990, South Korea invested 8.7 % during 1960–1990 and China invested about 8 % between 2003 and 2004. Development experience also suggests a lower level of infrastructure investment, 7 % of GDP, in order to maintain high economic growth (Nguyen and Dapice 2009). However, the progress of infrastructure development in Vietnam remains slower than other regional countries, including its neighbour and rival, China. Vietnam's infrastructure ranked 111th out of 133 countries surveyed in

Table 6.2 Comparison of quality of overall infrastructure of Vietnam and some neighbouring countries

Country	Rank/133	Score[a]
Singapore	2	6.7
Hong Kong SAR, China	3	6.7
Taiwan, China	19	5.8
Malaysia	27	5.4
Thailand	41	4.8
China	66	4.0
Indonesia	96	3.1
Philippines	98	3.1
Vietnam	111	2.8

[a] 1 = extremely underdeveloped; 7 = extensive and efficient by international standards. *Source* Schwab (2009)

Table 6.3 Rankings of infrastructure quality of Vietnam

Indicator	Rank/133
Quality of overall infrastructure	111 …x
Quality of roads	102 …x
Quality of railroad infrastructure	58 …x
Quality of port infrastructure	99 …x
Quality of air transport infrastructure	84 …x
Available seat kilometres	38 …√
Quality of electricity supply	103 …x
Telephone lines	36 …√

Note: x—competitive disadvantage; √—competitive advantage. *Source* Schwab (2009)

the global competitiveness report (GCR) 2009–2010 (Schwab 2009) (see Table 6.2). According to the GCR 2009–2010 (Schwab 2009), rankings of the quality of roads, ports and air transport infrastructure and electricity supply of Vietnam were among the lowest in the world and regarded as Vietnam's competitive disadvantages (see Table 6.3).

From 2006 to 2010, although the cost to export from Vietnam has been lower than in Thailand, Indonesia and the Philippines, it has increased faster, except for Indonesia. The time to export is also longest in Vietnam as shown in Table 6.4. Share of logistics costs was about 20–25 % of Vietnam's GDP in 2009, which was far higher than that of developed countries and even higher than its neighbour and rival, China (Manila Bulletin 2009). Costs of exporting and importing a standardized cargo of goods (including inland transport, and port and terminal handling) in Vietnam have seen an upward trend (see Figs. 6.8 and 6.9).

Inland transport of Vietnam is still slow and difficult. Road, the dominant means of transport, accounts for 67 % of cargo moved. Vietnam has about 175,000 km of roads, of which only about 10 % are asphalted, and only 15 % of bridges have a capacity of more than 10 tones (Vietnam Financial Review 2010). Almost all roads are one and two-lane roads with narrow widths, poorly designed junctions and restrictive vehicle weight limits. Railway transport is the slowest. Vietnam has a total of 2,600 km of rail lines connecting major population centres (Moore et al. 2010). However, most railways have been built almost 100 years ago and are in

Table 6.4 Time to export and to import of Vietnam and some neighbouring countries

Country	Time to export (days)					Time to import (days)				
	2006	2007	2008	2009	2010	2006	2007	2008	2009	2010
Singapore	5	5	5	5	5	4	4	4	4	4
Malaysia	18	18	18	18	18	14	14	14	14	14
Thailand	24	24	17	14	14	22	22	14	13	13
China	18	21	21	21	21	24	24	24	24	24
Indonesia	25	25	21	21	21	27	27	27	27	27
Philippines	17	17	17	16	16	18	18	18	16	16
Vietnam	24	24	24	24	22	23	23	23	23	21

Source Doing Business (2010)

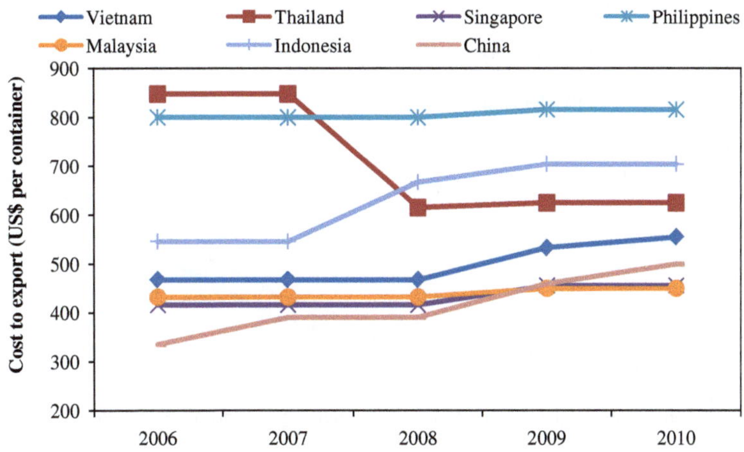

Fig. 6.8 Costs to export in Vietnam and some neighbouring countries. *Source* Doing Business (2010)

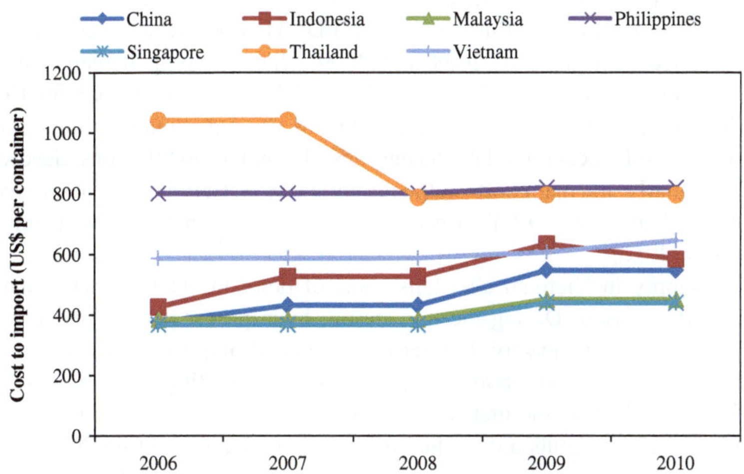

Fig. 6.9 Costs to import in Vietnam and some neighbouring countries. *Source* Doing Business (2010)

need of repairs (Davis Langdon and Seah International 1995). Trains can travel at a maximum speed of 70 km/h, but frequently have to slow to under 30 km/h on older sections of track and aged bridges along the network. Subway or elevated rail lines have not been built in urban centres of Hanoi and Ho Chi Minh City (Moore et al. 2010).

Vietnam has more than 3,000 km of coastline along the international seaways, which is considered a favourable condition for Vietnam to gravitate towards

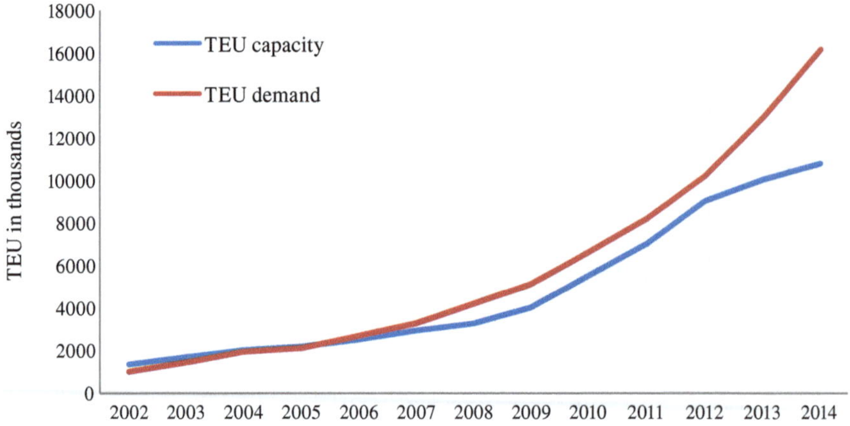

Fig. 6.10 Supply and demand at Ho Chi Minh City ports. Note: TEU—Twenty-foot equivalent units. *Source* American Chamber of Commerce in Vietnam (2006)

international trade. However, port congestion is threatening export growth of Vietnam. As noted by the American Chamber of Commerce in Vietnam (2006), port congestion in Ho Chi Minh city, where FDI has been most concentrated, was then forecasted to worsen in 2008 and 2009 when shipping demand began to exceed terminal capacities in 2007 (see Fig. 6.10). Ports in and around Ho Chi Minh City in the south accounting for about 65 % of total container throughputs in Vietnam can only accommodate feeder vessels under 30,000 tons deadweight (DWT). Similarly, the main ports in the northern city of Hai Phong can only receive ships of less than 30,000 DWT (America Chamber of Commerce in Vietnam 2006; Moore et al. 2010).

Power supply in Vietnam also falls short of demand. This primarily happens during the dry season. During the dry season, hydropower plants operate at only 40–50 % of designed capacity. Furthermore, reduced outputs at various plants for maintenance and servicing purposes are common, resulting in shortages to the nationwide grid. During the first seven months of 2008, there were 3,000 blackout incidents, i.e. 14 blackouts a day, due to system overloading. From 2001 to 2008, total electricity consumption grew at an annual average rate of 14.5 %. Electricity demand was forecasted to continue to grow at 14–15 % per year until 2015 (Nguyen and Dapice 2009; Moore et al. 2010).

The weaknesses of infrastructure in Vietnam are holding back the country from further economic development. Many international surveys have identified Vietnam's infrastructure as the biggest hindrance to the economic development process (Nguyen and Dapice 2009). According to various GCRs published by the World Economic Forum (Schwab 2008, 2009), ranking of the overall global competitiveness index (GCI) of Vietnam fell from 70/134 in 2008–2009 to 75/133 in 2009–2010. In the GCR 2009–2010, among the major competitiveness indicators, Vietnam's infrastructure remained the biggest drag on the country's further economic development (see Fig. 6.11).

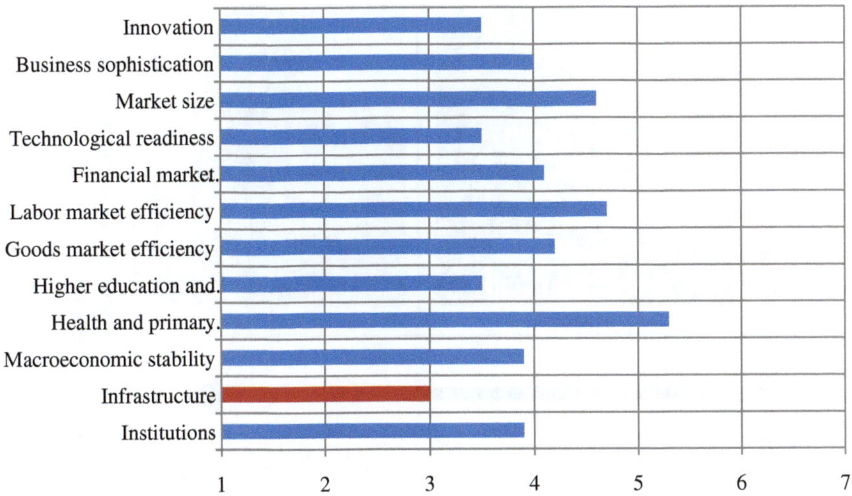

Fig. 6.11 Global competitiveness score of Vietnam. *Source* Schwab (2009)

Similarly, in other recent surveys such as those conducted by the Vietnam Business Forum and the Japanese External Trade Organization (as cited by Moore et al. 2010 and Nguyen and Dapice 2009), poor infrastructure, especially roads, ports and power system, was identified as the largest bottleneck for doing business in Vietnam. Weak infrastructure is therefore holding back the country to compete for both domestic and foreign investments in manufacturing and exports even though Vietnam is still a globally competitive, low wage manufacturer and commodity producer (Tran 2009; Vo and Nguyen 2009).

Vietnam's poor infrastructure is therefore regarded as one of the major constraints to further development of trade and international economic integration of the country. Limited state budget and ODA funds have been identified in many studies as the major cause for this poor progress. Solutions to the infrastructure bottlenecks in Vietnam seem to lie in further improvement in the PPP regulatory regime and the encouragement of private-sector participation, especially the foreign sector, in infrastructure delivery. Improvements in structuring the investments in infrastructure as a PPP could ease the problem of tight budget of the government with additional financing from the private sector (Warlters 2006; Vo 2007; Vo and Nguyen 2009; Moore et al. 2010). It is, however, notable that while investments in infrastructure development have recently increased, the quality of the infrastructure system of Vietnam has not improved very much. The country's experience in the construction of large-scale infrastructure projects has proved to be very problematic (see Fig. 6.12, for example) (as adapted from Thanh 2011; Tho 2008; Thinh 2010). The issue is further discussed in Sect. 6.5.6.

Fig. 6.12 Photographs of the poor infrastructure of Vietnam

6.4.2 Government Planning on Infrastructure Development

To deal with future infrastructure demand, the MPI, the Ministry of Transport (MoT) and the MoIT have developed numerous plans to invest in national and major municipal port, road, railway, airport and telecoms projects to 2020 (Prime Minister's Office of Vietnam, 2004a, b, 2005). Highlights of these plans consist of the following:

- Roads: to upgrade key segment of the North–South highway; to build a new highway along Vietnam's coastline; and to build new expressways connecting major business centres in northern, central and southern Vietnam;
- Railways: to consider a North–South high-speed rail line; to upgrade key rail lines; and to build new high-speed rail lines connecting major business centres in northern, central and southern Vietnam;
- Ports: to upgrade and expand major national ports; to build new deep water ports in northern, central and southern Vietnam, and new international terminal ports;

- Airports: to upgrade and expand the existing international airports and to build new international airports with capacities and standards comparable with regional benchmarks; to upgrade domestic airports with international standards;
- Electricity: to build new thermal power plants to reduce the reliance on hydropower; to build new hydropower plants that can facilitate water supply, drainage and flood control; to build the first nuclear power plant by 2020; to develop power plants using renewable energy; and to develop 200 and 500 kV transmission networks; and
- Telecoms: to develop a national information infrastructure in order to deploy high-capacity communications and innovative technology throughout the entire country, including remote areas; to increase telephone and Internet penetration rates and to ensure all provinces and cities to connect to broadband networks via optic fibre cable, TV cable and satellite as a basis for the application of information technology in e-trade, e-government, public services and other areas.

To meet the expected demand, Vietnam was estimated to need around USD 139 billion invested in infrastructure over the next five years (Moore et al. 2010). The major source of finance for infrastructure projects in Vietnam has been the state budget, including investments by major state-owned enterprises. In addition to government funding, ODA has accounted for a major portion of spending on infrastructure projects. Major donors of ODA are the World Bank, Japan and the Asian Development Bank (ADB) (Nam 2009). However, since Vietnam has been moving to a middle-income country, some ODA donors are planning to stop provision of ODA funds to Vietnam in the near future (Intellasia News Online 2011).

With limited state budget and constraints in getting ODA funds, the government has issued and introduced policy and regulatory documents to facilitate domestic private and foreign investments in the infrastructure sector (see Sect. 6.5.3). These measures are also expected to introduce competition to the infrastructure sector which has been monopolized by a single state-owned enterprise.

6.5 Construction Industry

6.5.1 Construction and Economic Growth

Since the economy started to grow rapidly in 1992, the construction industry of Vietnam has always been among the fastest-growing sectors from the demand side, but also the hardest hit during the 1997–1998 Asian financial crisis and the 2008 global financial crisis (see Fig. 6.13). In terms of total value added, the average growth of the sector was about 11 % per year during 1992–2008, while that of the economy was 7.7 % per year (United Nations Statistics Division 2010).

As shown in Fig. 6.14, before 1991, the contribution of the construction industry to GDP (the ratio of value added in construction to total value added of the economy) in Vietnam stayed almost the same at about 3.84 %. However, since

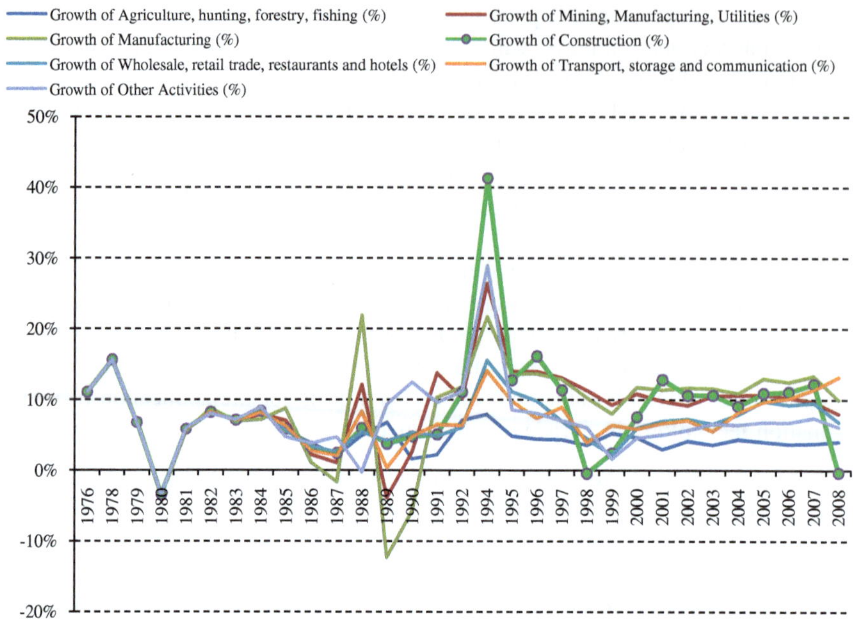

Fig. 6.13 Vietnam's growth by economic activities. Note: Data measured at constant prices 1990 US$. *Source* Data compiled from the United Nations Statistics Division (2010)

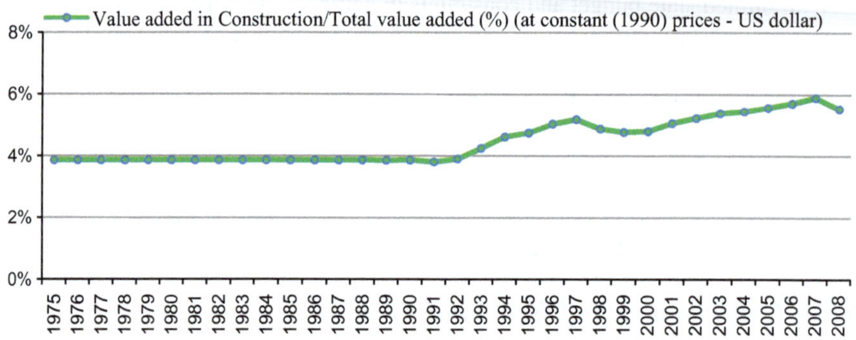

Fig. 6.14 Contribution of Vietnam's construction industry to GDP. *Source* Data compiled from the United Nations Statistics Division (2010)

1992, the ratio steadily increased from 3.88 % in 1992 to 5.17 % in 1997. Seriously affected by the Asian financial crisis, the ratio in 1998, 1999 and 2000 went down to 4.86, 4.75 and 4.78 %, respectively. However, since 2001, the contribution of the construction industry to the economy started to rise again from 5.05 % in 2001 to nearly 6 % in 2007, before going down to 5.53 % in 2008 due to the negative effects from the 2008 global financial crisis (United Nations Statistics Division 2010).

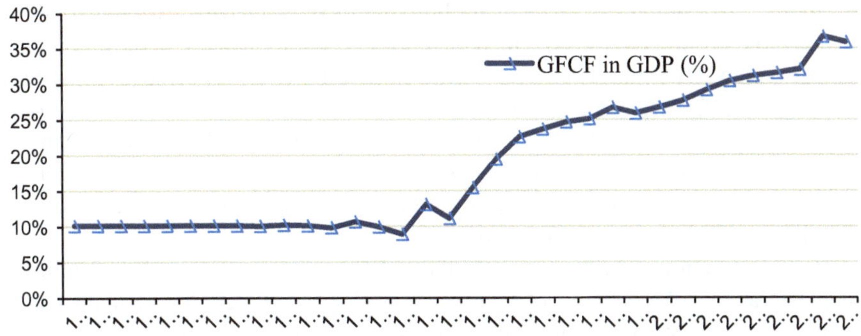

Fig. 6.15 Vietnam's GFCF in GDP. *Source* Data compiled from the United Nations Statistics Division (2010)

From the demand side, although published records of the contributions of the construction sector in GFCF are not available, the data of GFCF can be compiled from UNSD. Before 1990, GFCF comprised only about 10 % of GDP. Since 1991, the ratio increased steadily from 11 % in 1991 to 36 % in 2008 (see Fig. 6.15).

According to Business Monitor International (2010), infrastructure demand dominates the construction sector of Vietnam, accounting for about 42 and 46 % of total construction value in 2007 and 2008, respectively. With poor infrastructure provision, it was forecasted that infrastructure will account for an average of about 47 % in total construction value per year during the period 2009–2014 (Business Monitor International 2010).

During the period 1995–2008, employment in the construction industry has increased significantly from 793,000 people in 1995 to about 2.5 million people in 2008. However, employment in construction made up only a small share in total employment in Vietnam. In 2005, employment in construction accounted for about 4.7 %. The portion increased to 5.6 % in 2006. From 2005 to 2009, employment in construction grew about 7.6 % per year (Data compiled from the General Statistics Office of Vietnam (2010) and Pham (2008)).

6.5.2 Construction Labour, Materials and Equipment

Vietnam is well endowed with natural resources for the manufacture of building materials. Most of the construction materials can be produced domestically. Before 1986, the building materials sector, however, was dependent much on imported materials. Since the market liberalization in 1986, investments from both the public and private, including foreign invested, sectors in new production lines have grown considerably. The expansion of the domestic production facilities has helped the sector provide sufficient quality and quantity of building materials for domestic consumption, and gradually reduced the reliance on imports. In 2000, the supply of some major building materials had met domestic demand and started to contribute to exports. Total outputs of some major building materials have grown remarkably. For example, from 1990 to 2007, there had been a 14-fold increase in the amount of cement output. Over the seven-year period from 2000 to 2007, cement output had almost tripled, from about 13.3 million tons (Mt) in 2000 to 36.8 Mt in 2007. Total outputs of some major building materials from 2000 to 2007 are shown in Table 6.5. As part of the government strategy, it is expected that the production capacity will be further expanded. Forecast of the expansion of some key building materials to 2020 is listed in Table 6.6.

The construction industry currently has a labour force of more than 2 million people (Vietnam Association of Construction Contractors 2011). Vietnam has a comprehensive system of education and training for construction professionals and workers. Major education and training programs in infrastructure construction in Vietnam are offered by public institutions, including:

- Hanoi Architectural University;
- Ho Chi Minh City University of Architecture;
- University of Civil Engineering;
- The Faculty of Civil Engineering of the University of Transport and Communication;
- The Faculty of Civil Engineering of Ho Chi Minh City University of Transport;
- Post and Telecommunications Institute of Technology;
- The Faculty of Hydraulic Engineering of the Vietnam Maritime University;
- Academy of Managers for Construction and Cities;
- Construction Technical Colleges No. 1, No. 2 and No. 3;
- Construction College of Mien Tay;
- Urban Construction College;
- Construction College of Nam Dinh;
- Technical College for Construction Mechanics No. 1; and
- Construction Polytechnic No. 4 (Ministry of Construction 2008).

Some construction companies also provide training programs. These public institutions and construction companies provide 44,000 skilled labours to the industry each year. With investments in construction growing at an annual growth rate of 15–20 %, the skilled labour available every year accounts for only 25 % of demand. The lack of skilled workers was particularly acute in large construction projects where unskilled workers were employed to carry out jobs requiring technical skills (Vietnam Association of Construction Contractors 2011). The lack of a skilled workforce was reflected in the irrational structure of the workforce in the construction industry of Vietnam. While the rational structure should be 1 engineer/4 technicians/10 workers, that of the construction industry of Vietnam was 1 engineer/1.3 technicians/3 workers (Pham 2008).

The shortage of well-trained manpower, including trained engineers, project managers and on-site construction staff has impaired the sector's labour productivity. A comparison with 1995 as the base year showed that the ratio of value added in construction (at constant (1994) prices) to labour employed by the construction sector of Vietnam raised by 18 % from 18.4 million dong/employee in

Table 6.5 Total outputs of major building materials in Vietnam from 2000 to 2007

Material	Unit	2000	2001	2002	2003	2004	2005	2006	2007	2007/2000 (times)
Cement	million tons	13.3	16.1	21.12	24.12	26.15	30.8	32.4	36.8	2.7
Ceramic, granite tiles	million m^2	42	67.83	96.33	110.2	120	170	180	230	5.5
Glass	million m^2	30.71	30.72	38.77	38.35	43.68	74.76	81.31	80	2.6
Bricks	million	9,087	10,300	11,995	14,101	15,991	18,128	21,293	22,000	2.4
Roofing materials	million m^2	38.04	53.04	67.04	71.42	76.74	92.22	99.58	101	2.7
Building stones	million m^3	22.17	30.9	36.72	53.25	55.13	70.8	79	88	3.9
Slates	million m^2	1.52	1.65	1.8	2.0	2.5	3.2	4.5	6.0	3.9
Sand	million m^3	33.27	43	50.09	53.21	58.1	66.4	73	80	2.4

Source Tran (2008)

Table 6.6 Forecasts of capacities and outputs of key building materials in Vietnam from 2010 to 2020

Material	Unit	2010		2015		2020	
		Capacity	Output	Capacity	Output	Capacity	Output
Cement	million tons	65.59	59.02	99.5	88.5	>112	112
Tiles	million m^2	275.4	206	335	302	460	414
Glass	million m^2	172.4	93	172.4	135	228	200.4
Bricks	million	27	25	35.5	32	46.5	42
Roofing materials	million m^2	140	126	190	171	248	224
Building stones	million m^3	115	104	164	148	226	204
Sand	million m^3	107	97	151	136	211	190

Source Prime Minister's Office of Vietnam (2008)

1995 to 21.8 million dong/employee in 2001. However, from 2002 to 2005, labour productivity went below the 1995 level. The lowest labour productivity was 1.62 million dong/employee in 2004 or a drop of 12 % compared to that in 1995. After 2005, there was a small progress in labour productivity. The productivity increased to 19.2 and 20.4 million dong/employee in 2006 and 2007, respectively. On average, labour productivity in the construction industry grew a mere 1.3 % annually during the period 1995–2007 (see Fig. 6.16).

While the labour productivity of the construction sector recorded a little progress from 1995 to 2007 as mentioned above, the average income of the labour force during the same period saw an upward trend. As shown in Fig. 6.17, the average income of the labour force grew faster than labour productivity. Average income of labour in 2007 rose by 84 % compared to that in 1995, almost double the 1995 level.

Similarly, data compiled from the General Statistics Office of Vietnam (2010) indicated that the average housing and material construction prices year on year

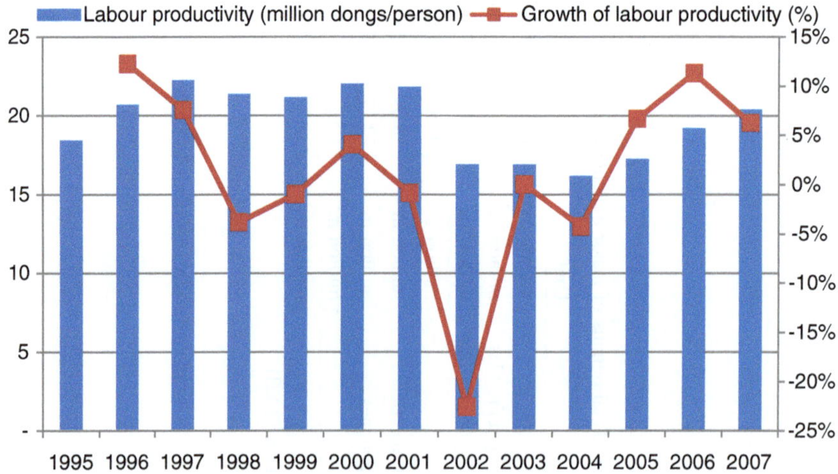

Fig. 6.16 Vietnam's labour productivity in construction industry. *Source* Data compiled from the General Statistics Office of Vietnam as cited by Pham (2008)

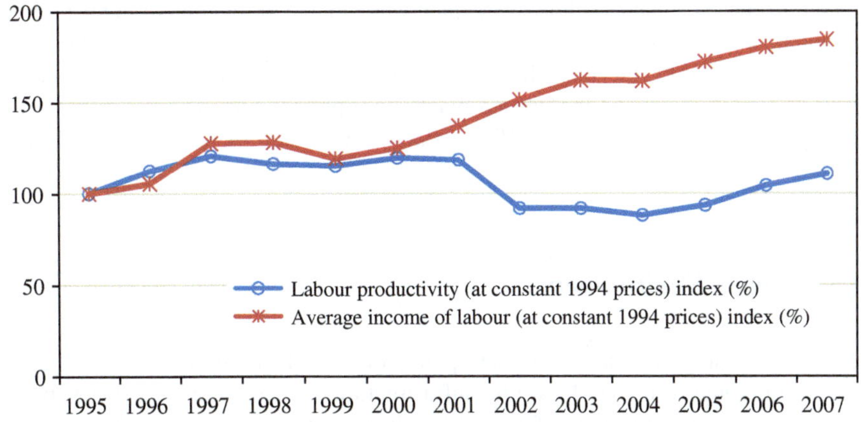

Fig. 6.17 Labour productivity and average income of labour in the construction industry of Vietnam. *Source* Data compiled from the General Statistics Office of Vietnam as cited by Pham (2008)

have grown faster recently. In 2000, average housing and material construction prices rose by only 4.5 %, while the prices climbed by 8 % in 2003, about 11 % in 2007, and more than 20 % in 2008, before going down to 3.5 % in 2009 due to the housing market's volatility (see Fig. 6.18).

Besides labour, current construction technologies and equipment were inadequate to meet the growing domestic demand, especially in large-scale projects such as large-scale bridge, tunnel and highway projects that required high quality and the use of modern construction technologies (Davis Langdon and Seah International 1995; Pham 2008). Most equipment required for the construction of infrastructure projects was not produced in Vietnam and therefore demand for equipment manufactured overseas is rising. According to Mr. Vo Quang Diem, Vice Head of the Department of Building Materials of MoC, every year about USD 2–3 million were spent on importing construction equipment from Italy, Germany, Britain, Spain,

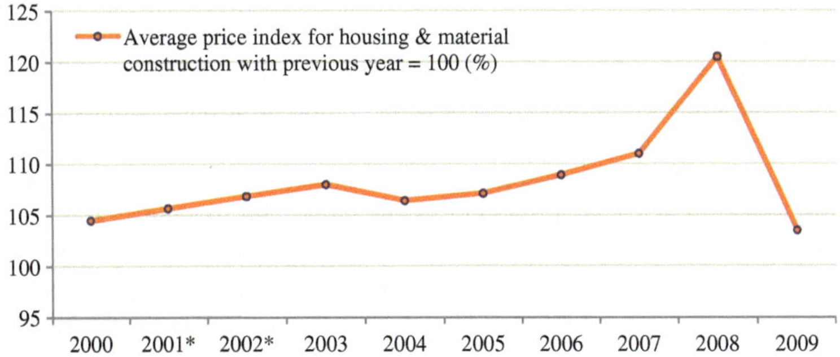

Fig. 6.18 Average price index for housing and material construction in Vietnam. *Source* data compiled from the General Statistics Office of Vietnam (2010)

China, Korea and Japan. Although Vietnam was capable of manufacturing partial or complete construction equipment, the market was overlooked by Vietnamese companies. Notably, the plant hire market was also depressed. Many large-scale projects tended to import the needed plants on a project-specific basis rather than through hire (May Xay Dung 2010; Vietnam Association of Mechanical Industry 2010c).

6.5.3 Legal Environment

Since 1986, following the government policy on liberalization of the market, legal and institutional framework of the infrastructure construction sector has been improved considerably. One of the initial changes was the issue of Decree 385/HDBT by the Ministers' Committee (HDBT) in 1990 to convert the whole development process of state investments towards a socialist-oriented market mechanism. Enclosed in the decree was a circular promulgating the bidding procedures for construction works. Prior to the issue of the decree, under the centrally planned system, a majority of public works were allocated to the public sector contractors. They did not need to go through any bidding procedures. The issue of the decree was therefore considered as a breakthrough step in raising the competitiveness of contracting firms in the country (Ministry of Construction 2008).

Further improvements in the legal environment and operating conditions for infrastructure construction have been achieved over the years. Due to the enormous need for infrastructural development in Vietnam, since 1993, the government has issued a number of regulations accommodating the build–operate–transfer contract and the like as authorized forms of investments. A government decree on the public–private partnership (PPP) scheme was also under consideration and a pilot PPP financing regulation was issued in late 2010 (Ministry of Construction 2008; Moore et al. 2010).

In addition, the introduction of Construction Law in 2003 is considered the most important milestone in the administration reform of the construction industry, which creates a favourable legal environment for investments from all economic sectors to meet the increasing construction demand. Following the issue of the Construction Law, the Tendering Law was introduced in 2005, promulgating the procedures for tendering for all types of construction projects. The introductions of the Construction Law and Tendering Law have officially ended the centrally planned and subsidy-based system in the construction sector (Ministry of Construction 2008). In 2010, as a measure to encourage the participation of domestic contractors as well as equipment and material suppliers, new regulations under Directive 494/CT-TTg were issued. These regulations require that projects financed by state capital [except for ODA (Official Development Assistance) funded projects] can only issue tenders for goods, equipment and EPC contracts to international suppliers if it is determined that local bidders do not meet the required standards (Davis Langdon and Seah International 1995; Moore et al. 2010).

At the same time, a number of appropriate standards, codes and regulations have also been issued and updated to improve practices and procedures in the development process of construction works in general and infrastructure construction in particular. Construction quality was mainly regulated through a set of construction standards (Ministry of Construction 2008). Since Vietnamese construction standards were often regarded as incomplete and out-of-date, application of other relevant technical standards such as ISO, British, American, French, Japanese standards and so on are allowed in Vietnam. These foreign standards were studied and modified to fit the conditions of Vietnam. At the same time, standards with reference to the local environment such as those related to soil and hydrology, fire, explosion, climate, safety, sanitation and so on have been imposed as well. Other regulations tackling violations of environmental and planning requirements and the applications of Vietnamese Codes have also helped to improve the industry's standard practices. Quality inspection was strengthened with the establishment of quality inspection units at both ministry level and provincial level (Chiang et al. 2004).

6.5.4 Industry Structure for Infrastructure Development

The importance of infrastructure has long been established in Vietnam. The central government has legislated the creation of infrastructure development implementation organizations since the post-war recovery period. After the reunification in 1975, most roads and communication networks in the whole country were damaged and outdated and could not catch up with the national economic growth. The need for the recovery of transport and communication networks was emphasized during the Fourth Party Congress in 1976. As a result, hundreds of public enterprises in transport and communication sectors in the whole country were created during the first decade after the reunification (Ministry of Transport 2008).

Following the liberalization in the mid- to late 1990s, the government decided to reform these public enterprises into public limited corporations. Their funding has been mainly drawn from the state budget and ODA resources. However, the state budget and ODA resources have shrunk as the country moved forward. To ease the budget constraints, the reform was speeded up since 2001. These corporations were restructured into a parent-subsidiary consortium comprising a management board to manage the parent and the creation of separate subsidiaries. To take up the burden for financing and constructing infrastructure, these public corporations could issue their own bonds, borrow from the private sector, or initiate a joint-venture with private companies (Vietnam Business Forum 2011). Major players engaged in the development of the transport and telecoms sector in Vietnam are discussed below.

The main developer of new roadways in Vietnam is the Vietnam Expressway Corporation (VEC). VEC is a state-owned corporation established in 2004 under the administration of the MoT. The Corporation has been designated to cover investments, development, management and maintenance for the national expressways. A number of key expressways have been developed by VEC, including key routes in the North–South link and in the Greater Mekong Subregion transport corridor (Vietnam Expressway Corporation 2009; Moore et al. 2010).

The Vietnam Railways Corporation (VNR) operates all the railway tracks and rolling stocks in Vietnam. The Corporation was established in 2003. The Corporation was designed to become a leading state-owned economic group in the country and the region with business areas focusing on railway transportation and operations for the economic development of Vietnam. Major national and provincial rail projects have been assigned to VNR, including

- To upgrade and modernize the existing railways, especially those on the North–South link;
- To invest in and develop elevated rail lines in urban cities, new rail lines and stations to connect major business and industrial zones in the whole country, creating a modern railway network; and
- To study and initiate the development of strategic railways, including the high-speed rail lines linking Hanoi and Ho Chi Minh City (Prime Minister's Office of Vietnam 2007a).

Similarly, the majority of airport upgrades and new terminal construction have been developed by Vietnamese aviation bodies including the Northern Airports Corporation, Middle Airports Corporation and Southern Airports Corporation under the administration of the Civil Aviation Administration of the MoT. These state-owned corporations were reformed from the airports authorities founded in 1993. The market for airports construction is dominated by Vietnamese state-owned companies. However, foreign contractors and designers, especially from Japan, have been involved in some investments financed by ODA (Moore et al. 2010).

Participation of foreign developers and contractors in the port sector is more considerable. Global port operators such as Hutchison from Hong Kong, SSA (Stevedoring Services of America) Marine from the USA and the Maritime and Port Authority of Singapore have participated in a series of projects, and often in the form of a joint-venture with major Vietnamese port operators such as Vietnam National Shipping Lines and Saigon Port. Main contractors and suppliers in these projects were from neighbouring countries, including China, Japan, Korea and Hong Kong. The Vietnam Waterway Construction Corporation and TediPort (a member of Tedi) under the MoT are the largest state-owned companies in port construction and design, respectively (Moore et al. 2010).

The Vietnam Posts and Telecommunications Group (VNPT) currently owns and controls the national communications network on behalf of the government. The group was reformed from the Vietnam Post and Telecommunication Corporation founded in 1995 under the Ministry of Information and Communication. The group currently accounts for approximately 94 % of the aggregate market, including operations in all telecom segments: equipment, engineering, construction and consulting (Nguyen et al. 2005).

In the energy sector, most large power projects were invested by the two state-owned groups including Electricity of Vietnam (EVN) and the Vietnam Oil and Gas Group (Petrovietnam). While EVN is the main builder of new power plants in the country, the role of Petrovietnam is increasingly important. Several foreign-invested projects were primarily those involved in renewable energy, offshore gas reserves and pipelines development.

Electricity of Vietnam was established in 1994 under the direction of the Ministry of Energy (now the Ministry of Industry and Trade) as the leading state-owned corporation in power supply. Major activities of the corporation involve power supply, investments and development in accordance with the state master plans and development plans of the power sector. The corporation was reformed into EVN, a state-owned economic group in 2006 with other corporations in the power sector as company members. EVN accounts for more than 90 % of total power generation with a transmission and distribution network to every district in the whole country (Prime Minister's Office of Vietnam 1994; Electricity of Vietnam 2008; Anh 2010).

The main contractors in infrastructure construction projects were state-owned construction companies. Major Vietnamese contractors and consultants are Construction Company #1, Lilama, Cienco, Licogi, Vinaconex and Tedi. There were also a few foreign contractors, who are mostly involved in ODA-funded projects. Most foreign companies participated in the sector as equipment and machinery suppliers rather than as engineering and construction contractors. However, foreign contractors have greater opportunities in large-scale infrastructure projects which require greater technical expertise or which are mainly funded by ODA sources. Many Japanese engineering and construction firms were awarded contracts in Japanese ODA projects in Vietnam. These foreign construction firms subsequently worked with local construction firms for project implementation (Moore et al. 2010).

More recently, the participation of foreign contractors, especially Chinese contractors, has increased. It was estimated that Chinese contractors was awarded 90 % of EPC contracts in Vietnam. Among them, many are major national projects in oil and gas, chemicals and power. Notably, Chinese contractors usually imported equipment, materials and even labour from China rather than worked with local firms for project implementation. Local firms were losing their domestic market. Government authorities have unofficially expressed cautious opinions about the level at which the sector should be opened to foreign participation, especially Chinese firms, and the real economic growth of Vietnam as well as the national security. It was argued that bidding requirements for large projects, which were often prepared by foreign consultants rigidly applying international formats, were not suitable in the Vietnam context. As a result, local bidders were rarely short-listed for these bids. In smaller projects, local contractors stood a greater chance. However, local bidders often had to deal with high costs in preparing bidding documents and kickbacks. As a result, low prices offered by Chinese contractors often helped them win the bids. In addition, the increase of non-compliance with the Tendering Law and exploiting some loopholes in the law had made the situation worse (Vietnam Association of Mechanical Industry 2010a, b).

6.5.5 National Professional Organizations

The Vietnam Federation of Civil Engineering Associations (VFCEA), formerly known as "the Vietnam Association of Building Technology", was founded in 1982. VFCEA is the largest professional organization in the sector. VFCEA is under the administration of the Vietnam Union of Science and Technology Associations (VUSTA) and a member of the Asian Civil Engineering Coordinating Council. The member list of VFCEA is shown in Table 6.7 (Vietnam Federation of Civil Engineering Associations 2011).

The Vietnam Bridge and Road Association (VBRA) founded in 1987 is a major national association in the infrastructure construction sector. VBRA is a member of VUSTA, the International Road Federation and the Road Engineering Association of Asia and Australasia. Members of VBRA are bridge and road associations of cities and central provinces, including the Ho Chi Minh City Bridge, Road and Port Association, and the Hanoi Bridge and Road Association (Vietnam Bridge and Road Association 2011). Vietnam Association of Construction Contractors (VACC) was founded in 1999. Members of the association are companies from all sectors of the construction industry. VACC officially became a member of the ASEAN Constructors Federation in 2002 (Vietnam Association of Construction Contractors 2010; Asean Constructors Federation 2011). These associations were created to provide opportunities for members to exchange ideas and innovations in construction technology, research and development. With members from all sectors of the industry, these associations have been playing an important role in shaping the future of the infrastructure construction sector.

Table 6.7 Member national associations of VFCEA

No.	National Associations
1	Vietnam Association of Port—Waterway—Offshore Engineering
2	Vietnam Society for Soil Mechanics and Geotechnical Engineering
3	Vietnam Association of Construction Economics
4	Vietnam Association of Structural Engineering and Construction Technology
5	Vietnam Urban Environment and Industry Zone Association
6	Vietnam Association of Construction Environment
7	Vietnam Association for Information Processing
8	Vietnam Lighting Association
9	Vietnam National Commission on Large Dams and Water Resources Development
10	Vietnam Association for Building Materials
11	Vietnam Aviation Construction Association
12	Vietnam Concrete Association

Source Vietnam Federation of Civil Engineering Associations (2011)

6.5.6 *Infrastructure Construction*

6.5.6.1 Infrastructure Investment Planning

There are a number of issues related to inappropriate master planning in project selection and investment coordination (Nguyen and Dapice 2009; Vo and Nguyen 2009). Investments in infrastructure in rapidly growing and key regions should be the most essential. However, many large-scale roads, ports and airports have been planned and built without regard to these regions. For example, in 2000, the Ho Chi Minh National Highway, built parallel to National Highway No. 1A (also known as National Highway No. 1) linking the north and the south, was planned to reduce the heavy traffic in National Highway No. 1A and thus to stimulate economic growth in the poorer regions in the highlands that it cut through. The construction work was divided into three stages with total expected cost of USD 1 billion. The first stage of the construction work was completed with an actual cost of more than USD 2 billion. Although many parts of the highway have been opened for use, traffic on these parts has not considerably increased and most vehicles still chose to travel on National Highway No. 1A. Da Nang port in central Vietnam was another example. It was built in the early 2000s and was ranked first class in Vietnam. However, it costs more and took more time to export goods through Da Nang port than through Sai Gon port in the south where economic activities were heavily concentrated (Nguyen and Dapice 2009). Consequently, these newly built infrastructures were under-utilized, while many rapidly growing regions were still facing severe infrastructure bottlenecks.

Similarly, the railway system of Vietnam, which has been in operation since 1936 (Vietnam Railways 2012), is in need of repair. However, annual capital expenditure on the sector remained insignificant that could not meet 40 % of minimal demand (Ho 2011). Uneven distribution of funds also occurred in the energy sector. While Vietnam was always short of electricity supply, the Electricity of Vietnam (EVN), the largest state-owned corporation in electricity supply, was expanding its investments in mobile phone and financial services as well as real estate developments (Nguyen and Dapice 2009).

Besides inappropriate planning, the report by the Standing Committee of the National Assembly of Vietnam on the implementation of the state-funded infrastructure construction policy from 2005 to 2007 indicated that the disjointed distribution of investments was common and showed an upward trend. A number of investments were planned individually albeit their overlaps. The Vice Head of the Government Inspection Agency, Mai Quoc Binh, maintained that if more attentive initial planning was carried out, billions of dollar could have been saved (Phuong 2008).

6.5.6.2 Infrastructure Project Implementation

Over the past few years, the government of Vietnam has approved a number of infrastructure investment projects. However, the increasingly poor performance of existing civil engineering works is affecting the effectiveness of these investments. Many roads in Vietnam are still in bad condition, suffering from insufficient structure (such as poor drainage and inadequate foundations or supporting structures) and poor maintenance. Even some newly built major municipal roads and bridges also suffered severe quality problems. For example, shortly after being built or upgraded, a number of roads in Ho Chi Minh City were dug up again for the installation of underground utilities, causing further traffic congestion. Finishing work was done improperly, and potholes were common on these roads (Tuan 2010). The first phase of the ring road around the central city of Ha Tinh was completed in 2009 with a total investment of VND 300 billion (around USD 19 million at that time). However, the road has already exhibited a number of serious surface distortions (Van 2010a, b).

The construction of Can Tho Stayed-cable Bridge over Hau River connecting Can Tho City and Vinh Long province in the south of Vietnam started in 2004. The bridge, the longest stayed-cable bridge in Southeast Asia, was designed by a Japanese consultant and constructed by Vietnamese, Japanese and Chinese contractors. The USD 342.6 million investment (2001 exchange rate) was funded mainly by ODA from Japan and a small corresponding portion from the state budget of Vietnam. While still under construction in 2007, two side spans of the bridge collapsed (Nguyen 2007). Although it was planned to be completed in 2008, the incident subsequently delayed the completion until 2010 (Nguyen and Dapice 2009).

The Kim Lien Road Tunnel, one of the most modern road tunnels in Vietnam was built to reduce traffic congestion in one of the major traffic corridors of Hanoi. The VND 467 billion (around USD 28 million at that time) investment was funded mainly by ODA and a small corresponding portion from the state budget of Vietnam. The tunnel was designed and constructed by Japanese firms (Do 2010). The construction of the tunnel started in 2006 and was planned to be completed within 22 months, but was delayed by a year. Completed in 2009, the tunnel was severely flooded on the opening day in 2009 (Duc 2009). A few months after the opening, a number of leaks also appeared in the tunnel's connections (Tien and Huong 2009).

The Thu Thiem Tunnel crossing Sai Gon River in the south, the longest river tunnel in Southeast Asia, is the main part of the East–West highway and tunnel project that run through southern Ho Chi Minh City; the total estimated cost was about USD 583 million and was mainly funded by ODA from Japan and a small corresponding portion from the state budget of Vietnam (Prime Minister's Office of Vietnam 2000a). The six-lane road tunnel is 1490 m long designed and constructed by Japanese firms. The construction works started in 2005. In 2008, the four under-river tunnel sections had a number of cracks when the construction of these tunnel sections was almost complete. The tunnel was opened for use in 2011 (Kap 2008; Vu 2008; Quang 2011).

The feasibility and cost effectiveness of building major infrastructure projects were questionable as well. One of the most problematic issues in building municipal roads is site-clearance-related issues, which often remarkably raised total development time and costs of these roads. For example, the construction cost for the Nam Ky Khoi Nghia Road, which played a key role in the development of the urban transport axis of Ho Chi Minh city, was only VND 68 billion, but site clearance cost was VND 654 billion, almost 10 times higher (Vo 2008). Likewise, many ports have been built albeit in a very low cost-effective way. For example, in 2004, the first phase of the Cai Lan Port in the northern city of Hai Phong was completed and available for use with a total cost of VND 1,500 billion (around USD 96 million at that time) funded by ODA from Japan and was expected to be one of the central ports in the north to accommodate ships of up to 40,000 DWT. The port, however, could accommodate only 15–20,000 DWT ships. Consequently, container cargos carried by 40,000 DWT ships were trans-shipped to 15,000 DWT ships before being unloaded in the port. This subsequently increased the shipping costs (and time) by tens of millions of USD per year (Do 2006). Upgrades to the port, including the dredging operation for the passage leading to the port to accommodate ships of up to 40,000 DWT, is currently under development with an investment of USD 100 million (Moore et al. 2010).

A development of a deep water seaport at Lach Huyen, also near Haiphong, is under consideration. The unstable site of the port requires a multi-billion dollar investment, let alone other issues. However, the feasibility of the development, which should be studied carefully, has been studied and verified by the same company (Dao 2007).

A report by the Standing Committee of the National Assembly discovered that within 3 years from 2005 to 2007, the number of infrastructure developments with fraudulent practices, or with time delay, low quality, or low-cost effectiveness have gone up both by total amount and by percentage. The number was 1,882 projects, accounting for 14.57 % of the total number of projects in 2005; and 3,173 projects, accounting for 18.1 % of the total number of projects in 2006. The number continued to increase to 4,763 projects in 2007. According to the general secretary of VFCEA, Mr. Tran Ngoc Hung, time delay was, in fact, worse than what was reported; construction time in Vietnam was often 1.5–2 times longer than that in other regional countries; and an upward trend could also be found in loss and waste in construction. Unnecessary bureaucracy was one of the major causes of time delay. On average, total time from acquiring investment approval to the beginning of construction works was about 42 months for group-A projects, 29 months for group-B and 23 months for group-C. Some group-A developments took 4 years to clear the site and 5 to 7 years to complete the construction works (Vo 2008).

Long et al. (2004) conducted a survey of the issues relating to the poor performance of large construction projects in Vietnam. The survey's respondents were owners, designers/consultants and contractors/subcontractors. Top-ranked factors included incompetent designers and contractors; poor estimation and change

Table 6.8 Major factors explaining the poor performance of large construction projects in Vietnam

Factors	Problems
Incompetent designers and contractors	Inadequate project management assistance
	Impractical design
	Lack of involvement through project life
	Contractor's financial difficulties
	Incompetent project team; and
	Poor site management
Poor estimation and change management	Excessive contractors and/or subcontractors
	Inaccurate time estimating
	Inaccurate cost estimating; and
	Excessive change orders
Social and technological issues	Obsolete technology
	Bureaucracy; and
	Fraudulent practices and kickbacks
Site-related issues	Slow site clearance; and
	Unsatisfactory site compensation
Improper techniques/tools	Inadequate modern equipment; and
	Improper planning and scheduling

management; social and technological issues; site-related issues, and improper techniques and tools. The problems for these factors (Long et al. 2004) are listed in Table 6.8.

Similarly, Le-Hoai et al. (2008) conducted a survey on the cause of delay and cost overruns in large construction projects located in robust development economic zones in Vietnam. The study revealed a number of issues in the implementation of construction projects, which were often funded by the government. These issues included slow payment of completed works, poor contract management, obsolete or unsuitable construction methods and unforeseen site conditions; poor site management and supervision, slow information flow between parties and poor project management assistance; mistakes in design, design changes and additional works; shortages of materials, inaccurate estimates and price fluctuations; financial difficulties of contractors and owners; obstacles from government; and shortages of skilled workers. Besides delay and cost overruns, other consequences caused by these problems were project failure, reduction of profit margin, waste of resources and loss of sponsors' as well as citizens' trust in government-funded projects (Le-Hoai et al. 2008).

In a recent work by Ling and Hoang (2010), a number of risks related to the implementation of construction projects in Vietnam were also identified. The major risks were corruption, termination of public projects, bureaucratic administrative system, changing and inconsistent regulations, inadequate legal framework, and fluctuation of exchange, interest and inflation rates.

While investments in infrastructure have increased, the level of development and the rate of implementation of infrastructure construction have not been improved as expected. Vietnam was still suffering from more and more infrastructure weaknesses. Investments in infrastructure developments were often large and funded by state budget, loans and aids from international organizations. Low quality, time and cost effectiveness of infrastructure construction in Vietnam thus not only resulted in huge waste of resources and under-capacity performance of infrastructure in the short term but also severely affected the economic development in the long term.

6.6 Government Institutions Involved in the Development Planning Process

6.6.1 Overview of the Governmental System

The government of Vietnam has a relatively vertical structure. Members of the government include a prime minister, deputy prime ministers, ministers and heads of ministry-equivalent agencies. The Prime Minister of Vietnam leads a cabinet currently composed of five deputy prime ministers and the heads of twenty-one ministries and commissions. The Prime Minister also has the power to establish, dissolve or restructure public agencies, all to be confirmed by the National Assembly. The government is in charge of the supervision of public sector agencies at all levels. The government is in turn under the supervision of the National Assembly and reports to the National Assembly, the Standing Committee of the National Assembly and the President. The President is elected by the National Assembly, and the Prime Minister is nominated by the President and approved by the National Assembly. The President, in principle, functions as Head of State and the nominal commander of the armed forces and chairman of the Council on National Defense (Government Web Portal of Vietnam 2011).

Although the government is responsible for the overall management of the economy, the broad parameters of national policy are set by the Communist Party of Vietnam, the only party in power in the country since independence. The actual role of the Party in state management is opaque. However, the occupancy of all senior government positions by party members has ensured the primary adherence of the Politburo guidelines. Party cells function in parallel with state agencies at all levels (Arkadie and Mallon 2003).

The National Assembly is the highest organ of the state. Almost 90 % of the members of the National Assembly are party members. Elections for member are held every 5 years. Generally, the National Assembly has authority over scrutinizing legislation, reviewing government plans, budgets and implementation performance. Although decisions of the National Assembly are still subject to the direction of the Party, the role of the National Assembly has become increasingly important in overseeing all government functions in recent years. The National

Assembly is considered by the bureaucracy as a significant source of authority and a potential arena for independent criticisms of government performance. The Assembly meets twice a year, although its Standing Committee meets monthly and many members function full-time on various committees (Arkadie and Mallon 2003).

6.6.2 Coordination of Planning Across Ministries

6.6.2.1 Development Plans

Official planning documents of Vietnam have varying time frames, ranging from 1 to 10 years. The ten-year Socio-Economic Development Strategy (SEDS) was the main long-term planning documents formulated by the central government (Communist Party of Vietnam 2005). Commonly, these development plans are long on broad thrusts, but short on detail and concrete commitments. The SEDS generally consisted of an analytical report that assesses the current status of the economy, the progress achieved under the previous plan and the challenges faced by the economy; defined goals for a range of economic and social indicators; and prescribed the development approaches and strategies to achieve these goals. The ten-year SEDS set overall development direction of the country within the next ten years including strategic goals, development approaches and major instruments to achieve the goals (Ministry of Planning and Investment 2007). The five-year Socio-Economic Development Plan (SEDP) was formulated based on the national long-term development strategy with more concrete and disaggregated targets, which in turn provided the basis for the preparation and approval of annual plans. The ten-year SEDS and five-year SEDP also served as the basis for the formulation of sectoral and local development plans, including economic infrastructure development. Accompanying the five-year SEDP was a Public Investment Program (PIP). The PIP formed the basis for developing a framework of capital allocation among investments (Pincus and Nguyen 2004; Ministry of Planning and Investment 2007). The planning system is shown in Fig. 6.19.

6.6.2.2 Planning Process

Vietnam has a long tradition of planning. After independence, Vietnam ran a centrally planned economy. During the period of central planning, investment planning relied on command and control. Since the implementation of economic reforms in the late 1980s, the planning has relied on incentives and a general policy framework. The present investment planning process can be described as a centralized, but formally consultative process. In principle, decision making at all levels conforms to the consensual style, which engages a wide range of actors before decisions are finalized. However, the extent to which these consultations can influence the final

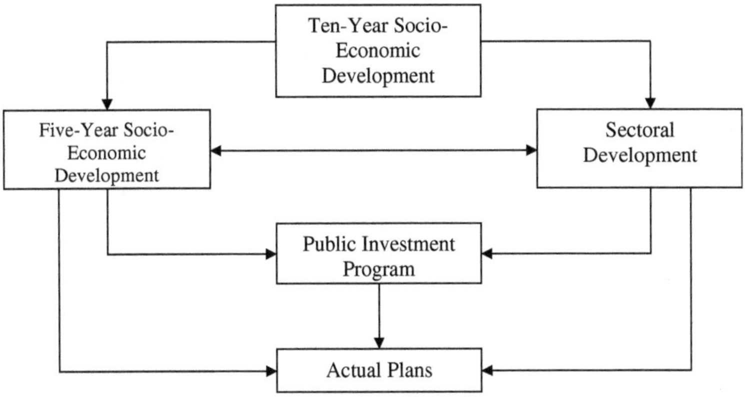

Fig. 6.19 Vietnam's national planning system

draft is not clear. The actual drafting process is complex and opaque to the general public and mass organizations at large (Pincus and Nguyen 2004).

The planning process is implemented in Vietnam at all level of government. While national and sectoral development plans are developed by the central government and relevant line ministries, local development plans are formulated by the local government. The planning process at central level (national and sectoral plans and strategies) is shown in Fig. 6.20. Proposals on SEDSs and SEDPs were prepared by the central government and submitted to the National Assembly for approval. The central government's preparation of SEDSs and SEDPs was led by the Prime Minister with the assistance of the MPI. The drafting process of the SEDSs for the next period started at the end of the previous period with a decision issued by the Politburo. Following the decision, the Prime Minister also established a strategic planning subcommittee composing of Party institutions, government agencies and national research institutes and a drafting committee led by the MPI (Arkadie and Mallon 2003; Pincus and Nguyen 2004; Ministry of Planning and Investment 2007).

The MPI was responsible for canvassing opinions from all the line ministries and other central government agencies before finalizing and delivering the national socio-economic development strategies and plans. More specifically, the Development Strategy Institute, a think-tank under the MPI was responsible for drafting the national plans based on sectoral strategies produced by line ministries and other central government agencies. The draft was then circulated for consultation with mass organizations and party institutions. During the final stages of consultation, the draft was published in national newspapers for public discussion. The final drafts were submitted to the National Assembly for ratification. The SDES and SEDP were then formally presented to the National Party Congress which is usually held every five years and approved by the National Assembly, following the resolution of the Party Congress (Arkadie and Mallon 2003; Pincus and Nguyen 2004; Ministry of Planning and Investment 2007).

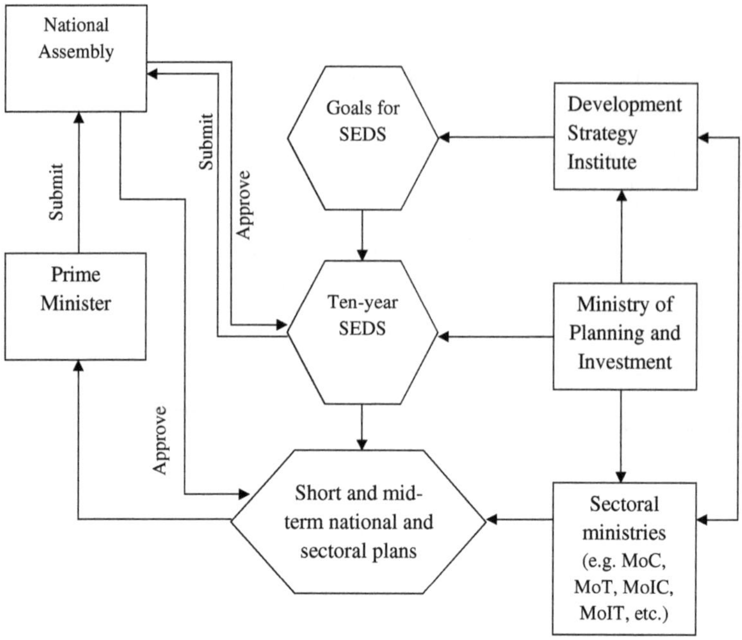

Fig. 6.20 Vietnam's planning process at central level

Based on these national development plans, master plans for sectoral development, such as those for construction, transportation, energy, telecommunication and trade were developed by line ministries and approved by the Prime Minister's Office. The consistency between sectoral plans and national socio-economic plans was coordinated by the MPI at the central level (Warlters 2006).

In relation to capital investments, the coordination was led by the MPI and the Ministry of Finance (MoF). While the MPI provided the Public Investment Program (PIP), including the list of investments, and investment budget compiled from sectoral master plans, the MoF was in charge of preparing the fiscal framework and recurrent expenditure for each sector. The PIP was prepared with reference to assessment of investment priorities given by the agencies involved. The PIP was finalized by the MPI and approved by the Prime Minister. Investments were classified into four levels (national, A, B and C) in terms of their scale and importance. National projects along with other special projects have to be approved by the Prime Minister following the resolution issued by the National Assembly. If necessary, the Prime Minister could set up a state committee to evaluate the investments which require approval by the Prime Minister, and the head of the state committee must be the minister of the MPI. Group-A projects were decided by relevant line ministries. Group-B and group-C investments were evaluated by relevant agencies at the provincial level (Prime Minister's Office of Vietnam 2009).

6.6.3 Ministries in Construction, Infrastructure and Trade Sectors

Various government bodies engage in trade, infrastructure and construction planning and policy-making. The roles and responsibilities of these organizations in the said activities are outlined in Table 6.9.

6.6.3.1 Ministry of Planning and Investment

The Ministry of Planning and Investment (MPI) was founded in 1995. The predecessor of the MPI was the National Planning Committee created in 1955. At the central level, the consistency between sectoral plans and national socio-economic plans was coordinated by the MPI. The departments involved in the said activities in infrastructure construction and trade planning are as follows:

Table 6.9 Ministries responsible for trade, infrastructure and construction sectors

Agency	Relevant responsibilities
Ministry of Planning and Investment	To ensure the consistency between sectoral development plans and national economic-social development plans; and
	To ensure sufficient and timely investments that are available for approved development plans
Ministry of Construction	To formulate policies and develop legislation and programs for building activities.
	To oversee the planning of urban areas, industrial zones, economic zones and hi-tech zones, and infrastructure facilities for these areas and zones
Ministry of Natural Resources and Environment	To formulate policies for and regulate land use, land price, land survey and mapping, land registration, land compensation and resettlement
Ministry of Transport	To formulate policies for and regulate the transport sector
	To develop strategies and plans of major transport subsectors, including transport infrastructure
Ministry of Information and Communications	To formulate policies for and regulate the telecom sector
	To develop strategies and plans of major telecom subsectors, including network infrastructure
Ministry of Industry and Trade	To formulate policies for and regulate the power sector
	To develop strategies and plans of major power subsectors, including power infrastructure
	To formulate policies for and regulate trading activities
	To develop strategies and plans of trade development

Source Ministry of Planning and Investment (2010); Ministry of Construction (2010); Ministry of Natural Resources and Environment (2010a); Ministry of Transport (2008); Ministry of Information and Communications (2010); Ministry of Industry and Trade (2010)

- Department for National Economic Issues;
- Department of Infrastructure and Urban Centers;
- Department of Investment Supervision and Appraisal;
- Department of Planning Management; and
- Public Procurement Agency (Ministry of Planning and Investment 2010).

These departments oversee the development strategies and master plans for infrastructure and cities and coordinate with other relevant units within the MPI to develop national and regional socio-economic development strategies.

6.6.3.2 Ministry of Construction (MoC)

The MoC was formed in 1973. However, its predecessor, the Ministry of Architecture had been established since 1958. The MoC was created in order to assume uniform state management of the construction industry in the whole country. The MoC is the state administration in charge of policy-making and regulatory matters in construction management and planning. More specifically, the major functions of the MoC are to create and implement policies, set construction standards, norms and rules, design processes and regulations, systems of limits, technical and economic criteria, limits of costs of investments and construction consultancy, and regulations for specialized construction works. The MoC also oversees state management of standards of quality, produces guidelines for the operation of construction engineering and consultancy enterprises, and implementation of the regulations on management of investment and construction. The MoC is also tasked to evaluate technical designs and budget for group-A investments and major national construction works. The MoC has 11 entities overseeing different areas of the construction industry, including:

- The Department of Architecture and Spatial Planning: The department oversees the planning of urban and rural areas, industrial zones, economic zones, hi-tech zones and important international-border gates.
- The Infrastructure Facility Agency: The agency oversees the planning of infrastructure facilities for urban transport infrastructure, water supply and drainage, lighting, greenery, cemetery, and waste treatment for urban areas, industrial zones, economic zones and hi-tech zones.
- Urban Development Agency: The agency oversees the planning of urban development.
- The Housing and Property Market Management Agency: The agency oversees housing development and the property market.
- The Department of Building Materials: The department oversees the planning of exploration, exploitation of minerals for the manufacture of building materials, and the demand and supply of building materials.
- The Department of Construction Economics: the department oversees the preparation of economic criteria, costs of investments and construction consultancy, and building material pricing.

- The Department of Science, Technology and Environment: The department is incharge of research and development of science and technology; advanced technology application; technology transfer; goods standards, measurements and quality; industrial properties; technological barriers in trade; environment protection; environment and science-technology information.
- The State Authority for Construction Quality Inspection: The authority is incharge of quality management for and inspections of construction works.
- The Department of Investment and Construction Management: The department is incharge of the development process in construction, and the operation of construction firms from all economic sectors.
- The Department of Legal Affairs: The department oversees the building of legal framework and related matters in areas within the Ministry's functions.
- The Department of Finance and Planning: The department is tasked, inter alia:
  - to compile strategies, master plans, long-term plans, five-year plans and annual plans for the development of the construction industry, as well as development plans for the Ministry's subordinate bodies; and
  - to compile and provide feedbacks on subjects related to construction and the implementation of development strategies and plans for the construction sector, which are part of national, regional socio-economic development plans or of master plans of other sectors.

Besides, the MoC has three research institutes involved in studying and reviewing the development of the construction sector, including Vietnam Institute of Architecture, Urban and Rural Planning; Institute of Construction Economics; and Vietnam Institute for Building Science and Technology (Prime Minister's Office of Vietnam 2008b; Ministry of Construction 2010).

6.6.3.3 Ministry of Natural Resources and Environment

The Ministry of Natural Resources and Environment was officially founded in 2002. The Ministry was formed by uniting the General Directorate of Land Registry, General Directorate of Hydrometeorology, Bureau of Environment (under the Ministry of Science, Technology and Environment), Bureau of Geology and Mineral Resources (under the Ministry of Industry), and Water Resource Management Department (under the Ministry of Agriculture and Rural Development) (Ministry of Natural Resources and Environment 2010a). Relating to land, major tasks of the General Directorate of Land Registry under the Ministry include taking a lead in implementing and monitoring long-term strategies for land use approved by the Prime Minister's Office; directing the process of national and local land use planning; coordinating with other relevant ministries to provide land use price frames and regulate land price-related issues; land surveying and mapping; land registration and certification; guiding and monitoring land compensation, resettlement and assistance; directing and coordinating with other relevant agencies to regulate and monitor land acquisition, transfer of land use titles and land optimization in accordance with land use master plans (Ministry of Natural Resources and Environment 2010b).

6.6.3.4 Ministry of Transport

The predecessor of the Ministry of Transport (MoT), the Ministry of Public Transport, was established since independence in 1945. In 1955, the Ministry of Public Transport was renamed as the Ministry of Post and Transport to cover the postal sector. However, the Ministry was renamed as the Ministry of Transport after the postal sector was separated from the Ministry's functions in 1992. The MoT engages in state management of roads, railway, inland waterway, maritime and airway transportation for the whole country. The development of major transportation sub-sectors is administered by separate agencies, including:

- The Department of Planning and Investment;
- The Directorate for Roads of Vietnam;
- Vietnam Railway Administration;
- Vietnam Inland Waterways Administration;
- Vietnam Maritime Administration;
- The Civil Aviation Administration of Vietnam; and
- Transport Engineering Construction and Quality Management Bureau.

Strategies, master plans, long-term plans, five-year plans, annual plans, national programs and projects for the development of major transportation subsectors, including transport infrastructure, are studied and prepared by these administrations. The Department of Planning and Investment under MoT is responsible for compiling these plans as well as evaluating plans and investments in transport infrastructure within the Ministry's authority. The MoT also has two institutes involved in the formulation of development strategy for transport development in Vietnam, namely the Transport Development Strategy Institute and the Institute of Transport Science and Technology (Ministry of Transport 2008).

6.6.3.5 Ministry of Information and Communications

The communication sector was under the administration of the Ministry of Posts and Transport founded in 1955. In 2002, the Ministry of Posts and Telecommunications (MPT) was founded and became the primary driver of telecom strategies, policies and regulations. However, since 2007, the MPT was reformed into the Ministry of Information and Communications (MoIC) to include the state management of the press and publication-related issues, which used to be functions of the old Ministry of Culture and Information. Based on the recent changes, the MoIC is responsible for state administration relating to press; publications; posts and courier services; telecom and Internet; wave transmission and emission; television-frequency management; information technology and electronics; radio and television broadcasting; and national information infrastructure. Subordinate agencies were created to administer each subsector. The department in charge of the telecom sector is the Department of Telecommunications. More specifically, the department oversees policies, standards, strategies, master plans, development plans and programs in such areas as telecom, Internet, as well as wave transmission and emission. Planning of network expansion in telecom, Internet, wave transmission and emission are also part of the functions of the National Institute of Information and Communications Strategy, an independent unit of the MoIC (Ministry of Information and Communications 2010).

6.6.3.6 Ministry of Industry and Trade

The new Ministry of Industry and Trade (MoIT) was created in 2007. Its predecessor was the Ministry of Economics founded in 1945. The Ministry of Economics was renamed the Ministry of Industry and Trade in 1951. The then MoIT was subsequently divided into several ministries before the two separate ministries, the Ministry of Trade and the Ministry of Industry, were formed in 1997. The two ministries were merged again in 2007 to form the current MoIT.

Under the MoIT, the Energy Department is responsible for the planning of the power sector. The department is tasked, inter alia, to prepare strategies, development policies and national master plans for the development of oil and gas, power, new energy and renewable energy resources. The MoIT also created the Electricity Regulatory Authority to oversee pricing, new investments in generation and transmission, and gradual liberalization of the generation, wholesale and retail markets. In addition, the Institute of Energy under MoIT plays an important advisory role in the planning of energy and power development.

Apart from industrial activities, the MoIT sets policy and develops legislation, strategies, master plans and programs for trading activities, including export and import, market management, trade promotion, e-commerce, trading services, international trade integration, competition management, anti-monopoly controls, the application of safeguards, anti-dumping and anti-subsidy measures, and consumer protection. Separate departments were created to oversee each activity, including:

- The Department of Export and Import;
- The Department of International Cooperation;
- The Department of Multilateral Trade Policy;
- Vietnam Trade Promotion Agency;
- Vietnam Competition Authority;
- Market Control Department; and
- Vietnam e-Commerce and Information Technology Agency.

In addition, the MoIT has separate departments responsible for developing trade relations between Vietnam with specific regions, including the Asia-Pacific Market Department; the Europe Market Department; the America Market Department; and

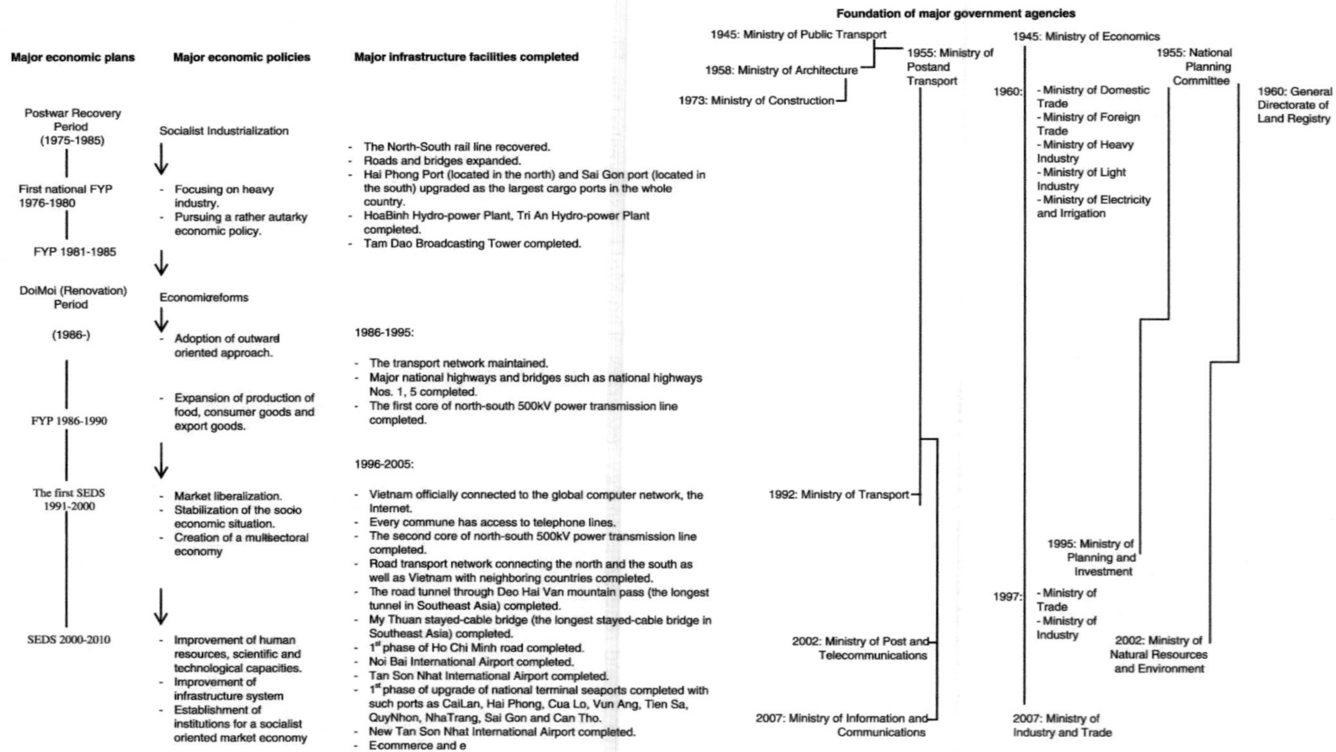

Fig. 6.21 Timeline of major economic policies, significant infrastructure development events and related institutional changes

the Africa, West Asia and South Asia Market Department. The Ministry's think-tank on trade matters is the Vietnam Institute for Trade (Prime Minister's Office of Vietnam 2007b; Ministry of Industry and Trade 2010).

6.7 Conclusions

Over the last 20 years, Vietnam has made significant development progress to become one of the fastest-growing economies in Asia. Vietnam's achievements over the past decades have been largely attributed to a combination of liberalization of markets, implementation of "open-door" policy and integration into the world economy, and a stable macroeconomic environment. Major economic policies that Vietnam has implemented are summarized in Fig. 6.21. The country has just entered into a lower-middle-income country status and preparing a new set of SEDS aiming to establish the foundation for Vietnam to become a modern, industrialized country by 2020 (Than and Tan 1993; Vo 2007; Asian Development Bank 2009; United Nations 2010b). However, the SEDS for 2001–2010 ended with many challenges.

While the importance of infrastructure financing has grown significantly, the efficiency in public investments in infrastructure is one of the key issues challenging the country. It was noticeable that investments in infrastructure development in Vietnam have heavily been increased over the past decades. The level of infrastructure development, however, has not been improved correspondingly. The infrastructure system in Vietnam is still regarded as one of the major factors holding back further development of trade and international economic integration of the country. The intriguing question is therefore why public investments in infrastructure have not been efficiently used. Further research should be conducted to find out what the problems affecting the efficient use of public investments in infrastructure to support trade and economic development in Vietnam.

Experience in Vietnam suggests that problems in the planning and execution of infrastructure construction would affect the efficient use of public investments in infrastructure. In Vietnam, a considerable portion of public investments goes to infrastructure construction. All major large-scale infrastructure projects are planned at central level. As reviewed earlier, the centralized planning process in Vietnam is complicated and opaque. Although the process is described as consultative, the actual participations of related actors and the influence of these consultations are not clear. From the literature review, poorly selected, uneven and disjointed infrastructure investments in Vietnam would therefore resulted from lack of transparency and accountability, institutional weaknesses in planning, politicized decision making as well as the inadequate capacity of estimating and monitoring rates of return. The efficient use of public investments in infrastructure could also be affected by a number of issues in the execution of infrastructure development plans. Reviews of construction sector in Vietnam have shown the low quality, time and cost effectiveness of infrastructure construction, which would result in the poor performance of infrastructure system and waste of national resources in the short term as well as affecting economic development in the long term. This suggests the efficiency factor in infrastructure construction cannot be neglected. From the literature review, issues that could affect the efficiency of the execution of government-funded infrastructure construction are corruption in infrastructure construction, shortage of local construction firms' capabilities and resources, land acquisition issues and institutional weaknesses in infrastructure construction.

References

Abbott, P., Bentzen, J., & Tarp, F. (2009). Trade and development: Lessons from Vietnam's past trade agreements. *World Development, 37*(2), 341–353.

America Chamber of Commerce in Vietnam. (2006). *Concerns over Vietnam's infrastructure development—Export container growth*. Ho Chi Minh City in Vietnam: America Chamber of Commerce.

Anh, B. (2010). EVN: To supply sufficient electricity from July. *VnEconomy*. Retrieved from http://vneconomy.vn/20100626103034502P0C19/evn-cap-dien-se-on-dinh-tu-thang-7-toi.htm.

Arkadie, B. V., & Mallon, R. (2003). *Vietnam: A transition tiger?*. Australia: Asia Pacific Press.

Asean Constructors Federation. (2011). Chronology event. Retrieved March 28, 2011, from http://www.aseanconstructorsfederation.org/federation_chronology.html.

Asian Development Bank. (2009). Asian Development Bank and Vietnam: Fact sheet.

Athukorala, P.-C. (2009). Economic transition and export performance in Vietnam. *ASEAN Economic Bulletin, 26*(1), 96–114.

Business Monitor International. (2010). *Vietnam infrastructure report Q4 2010, including 5-year industry forecasts*. London: Business Monitor International.

Chiang, Y.-H., Anson, M., & Raftery, J. (2004). *The construction sector in Asian economies*. New York: Spon Press.

Communist Party of Vietnam. (2005). *75 years of the communist party of Vietnam, 1930–2005: A selection of documents from nine party congresses* (1st ed.). Hanoi: The Gioi Publishers.

Dao, K. (2007, June). Construction of sea-port: Warnings before it's too late. *Thoi Bao Kinh Te Viet Nam*. Retrieved from http://www.hids.hochiminhcity.gov.vn/xemtin.asp?idcha=1681&cap=3&id=4238.

Das, S., & Shrestha, O. (2009). Vietnam: Further challenges in 2009. *ASEAN Economic Bulletin, 26*(1), 1–10.

Davis Langdon & Seah International. (1995). *Construction and development in Vietnam*. Hong Kong: Davis Langdon & Seah International.

Dinh, V. A. (2006). *Vietnam economy 2001–2005 and socio-economic development plan 2006–2010* (Central Institute for Economic Management Working Paper). Vietnam: Central Institute for Economic Management.

Do, H. L. (2006, January 1). Quang Ninh: 40,000 DWT ships cannot enter Cai Lan Port. *Tuoi Tre Online*. Retrieved from http://tuoitre.vn/Chinh-tri-Xa-hoi/117850/Quang-Ninh-Tau-4-van-tan-khong-the-vao%C2%A0cang-Cai-Lan.html.

Do, H. (2010). Disputes between parties relating to the Kim Lien road tunnel. *Doanh Nhan Viet Nam Toan Cau*. Retrieved from http://www.baomoi.com/Giai-quyet-tranh-chap-goi-thau-ham-duong-bo-Kim-Lien/148/4886058.epi.

Doing Business. (2010). Trading across borders. Retrieved December 28, 2010, from http://www.doingbusiness.org/data/exploretopics/trading-across-borders.

Duc, V. (2009, June 16). Kim Lien road tunnel: Flooded… on the opening day. *Voice of Vietnam News*. Retrieved from http://vovnews.vn/Home/Ham-duong-bo-Kim-Lien-Vua-thong-da-ngap/20096/114221.vov.

Electricity of Vietnam. (2008). About us. Retrieved March 20, 2011, from http://www.evn.com.vn/Default.aspx?tabid=60&TopicId=2&language=vi-VN.

General Statistics Office of Vietnam. (2010). Statistical data. Retrieved December 18, 2010, from http://www.gso.gov.vn/default_en.aspx?tabid=622.

Government Web Portal of Vietnam. (2010a). Some facts about Vietnam's geography (in Vietnamese). Retrieved December 24, 2010, from http://www.chinhphu.vn/cttdtcp/vi/nuocchxhcnvn/thongtintonghop/thongtintonghop_dialy.html.

Government Web Portal of Vietnam. (2010b). A summary of Vietnam's history (in Vietnamese). Retrieved December 25, 2010, from http://www.chinhphu.vn/cttdtcp/vi/nuocchxhcnvn/thongtintonghop/thongtintonghop_lichsu.html.

Government Web Portal of Vietnam. (2011). Members of the present government. Retrieved April 10, 2011, from http://www.chinhphu.vn/cttdtcp/vi/cpchxhcnvn/cpduongnhiem/cpduongnhiem.html.

Harvie, C., & Tran, V. H. (1997). *Vietnam's reforms and economic growth*. London: Macmillan.

Ho, T. (2011). Railway infrastructure: Too many shortcomings *Giao Thong Van Tai*. Retrieved from http://giaothongvantai.com.vn/kinh-te/du-an-dau-thau/201112/Cong-tac-GPMB-con-qua-nhieu-bat-cap-32953/.

Intellasia News Online. (2011). UK to stop ODA provision for Vietnam from 2016. *Intellasia News Online*. Retrieved from http://www.intellasia.net/news/articles/governance/111317320.shtml.

Kap, T. L. (2008). Thu Thiem tunnel: All four tunnel sections crack. *Tuoi Tre Online*. Retrieved from http://tuoitre.vn/Chinh-tri-Xa-hoi/274192/Ham-Thu-Thiem-Ca-4-dot-deu-nut.html.

Kien, K. (2010). Vietnam's development goals, 2011–2020. Vietnam communist party online newspaper. Retrieved from http://www.cpv.org.vn/cpv/Modules/News_English/News_Detail_E.aspx?CN_ID=396692&CO_ID=30113.

Le-Hoai, L., Lee, Y. D., & Lee, J. Y. (2008). Delay and cost overruns in Vietnam large construction projects: A comparison with other selected countries. *KSCE Journal of Civil Engineering, 12*(6), 367–377.

Ling, F. Y. Y., & Hoang, V. T. P. (2010). Political, economic and legal risks faced in international projects: Case study of Vietnam. *Journal of Professional Issues in Engineering Education and Practice, 136*(3), 156–164.

Long, N., Ogunlana, S., Quang, T., & Lam, K. (2004). Large construction projects in developing countries: A case study from Vietnam. *International Journal of Project Management, 22*(7), 553–561.

Manila Bulletin (2009, March 12). High logistics costs hold back Vietnam's economic development. *Manila Bulletin*. Retrieved from http://www.mb.com.ph/node/198686.

May Xay Dung (2010). Markets for construction plant and equipment overlooked. *May Xay Dung*. Retrieved from http://www.mayxaydung.vn/tintuc/3/90/thi-truong-may-xay-dung-bi-bo-trong.htm.

Ministry of Construction. (2008). *The history of construction industry of Vietnam*. Hanoi: Construction Publishing House.

Ministry of Construction. (2010). About us (in Vietnamese). Retrieved December 8, 2010, from http://www.xaydung.gov.vn/web/guest/gioi-thieu;jsessionid=02CA1415CC57A53E18CDC922AD80BBC5.

Ministry of Industry and Trade (2010). About us. Retrieved December 28, 2010, from http://www.moit.gov.vn/web/guest/gioithieu?timestamp=1302880149706.

Ministry of Information and Communications. (2010). About us. Retrieved December 28, 2010, from http://mic.gov.vn/gioithieu/Trang/Gi%E1%BB%9Bithi%E1%BB%87u.aspx.

Ministry of Natural Resources and Environment. (2010a). History of the Ministry. Retrieved December 29, 2010, from http://www.monre.gov.vn/v35/default.aspx?tabid=674&cateid=23.

Ministry of Natural Resources and Environment. (2010b). Organizational structure (in Vietnamese). Retrieved December 30, 2010, from http://www.monre.gov.vn/v35/default.aspx?tabid=674&cateid=18.

Ministry of Planning and Investment. (2007). About building development strategy. Retrieved December 20, 2010, from http://www.mpi.gov.vn/portal/page/portal/bkhdt/ptktxh/thptktxh/tctbvclpt?p_page_id=412546&p_cateid=353543&item_id=410714&article_details=1.

Ministry of Planning and Investment. (2010). About us. Retrieved April 8, 2011, from http://www.mpi.gov.vn/portal/page/portal/bkhdt/gtcbkhdt.

Ministry of Transport. (2008). About us. Retrieved March 25, 2011, from http://www.mt.gov.vn/Default.aspx?tabid=1&catid=100&articleid=25.

Moore, C., Nguyen, T. T. H., & Saladini, M. (2010). *Vietnam infrastructure*. Ho Chi Minh City: Italian Trade Commission and Mekong Research.

Nam, H. (2009). ODA for Vietnam in 2010 reaching a record high. *Voice of Vietnam News*. Retrieved from http://vovnews.vn/Home/ODA-cho-Viet-Nam-nam-2010-dat-ky-luc/200912/128340.vov.

National Assembly of Vietnam. (1997). *Commercial Law of Vietnam*. Retrieved from http://vanban.moit.gov.vn/search/View.aspx?ID=3675.

National Assembly of Vietnam. (2001). *Customs Law of Vietnam*. Retrieved from http://www.customs.gov.vn/Lists/VanBanPhapLuat/ViewDetails.aspx?ID=3944.

Nguyen, V. T. (2007). *Some comments on the Can Tho Bridge's quality incident*. Ho Chi Minh City: Ho-Chi-Minh-City Bridge and Road Association.

Nguyen, X. T., & Dapice, D. (2009). *Vietnam's infrastructure constraints* (Policy Dialogue Paper No. 3). Cambridge: ASH Institute for Democratic Governance and Innovation, John F. Kennedy School of Government.

Nguyen, T. H., Pham, Q. T., & Gullish, J. (2005). *Competition review of the Vietnamese telecom sector*. Hanoi: United States Agency for International Development.

Ohno, K. (2009). Avoiding the middle-income trap: Renovating industrial policy formulation in Vietnam. *ASEAN Economic Bulletin, 26*(1), 25–43.

Pham, V. B. (2008). *Improvement of the productivity of the construction sector in Vietnam*. Paper presented at the 14th Asia construct conference. Retrieved from http://www.asiaconst.com/past_conference/conference/14th/03Vietnam.pdf.

Phuong, A. (2008, September 23). The implementation of state-funded infrastructure construction: "old pain" persists. *Sai Gon Giai Phong Online*. Retrieved from http://www.sggp.org.vn/chinhtri/2008/9/166031/.

Pincus, J., & Nguyen, T. (2004). *Poverty reduction and strategy process and national development strategies Asia: Country study Vietnam*. London: University of London.

Prime Minister's Office of Vietnam. (1994). *Decision no. 562/1994/QĐ-TTg on the establishment of the Electricity Corporation of Vietnam*. Retrieved from http://www.chinhphu.vn/portal/page?_pageid=578,33345598&_dad=portal&_schema=PORTAL&docid=2269.

Prime Minister's Office of Vietnam. (2000a). *Decision no. 622/2000/QĐ-TTg on the approval of the East-West Highway and tunnel project in Ho Chi Minh City*. Retrieved from http://www.thuvienphapluat.vn/archive/Quyet-dinh/Quyet-dinh-622-QD-TTg-phe-duyet-dau-tu-Du-an-xay-dung-dai-lo-Dong-Tay-thanh-pho-Ho-Chi-Minh-vb17474t17.aspx.

Prime Minister's Office of Vietnam. (2000b). Instruction for the 2001–2010 export–import development strategy. Retrieved from http://vanban.moit.gov.vn/search/View.aspx?ID=4207.

Prime Minister's Office of Vietnam. (2004a). *Decision 206/2004/QĐ-TTg on development strategy for transportation to 2020*. Retrieved from http://www.chinhphu.vn/portal/page?_pageid=33,20664807&_dad=portal&_schema=PORTAL&p_cateid=20664306&item_id=20668872&vbpq_details=1.

Prime Minister's Office of Vietnam. (2004b). *Decision no. 176/2004/QĐ-TTg on development strategy for electricity from 2004 to 2010, and direction to 2020*. Retrieved from http://www.chinhphu.vn/portal/page?_pageid=33,177323&_dad=portal&_schema=PORTAL&p_cateid=178955&item_id=193112&vbpq_details=1.

Prime Minister's Office of Vietnam. (2005). *Decision no. 246/2005/QĐ-TTg on development strategy for information technology and communication of Vietnam to 2010 and direction to 2020*. Retrieved from http://www.chinhphu.vn/portal/page?_pageid=33,19067597&_dad=portal&_schema=PORTAL&p_cateid=19064848&item_id=19071390&vbpq_details=1.

Prime Minister's Office of Vietnam. (2007a). *Decision no. 621/2007/QĐ-TTg on the approval of the development plan of Vietnam Railways Corporation from 2007–2010*. Retrieved from http://www.thuvienphapluat.vn/archive/Quyet-dinh/Quyet-dinh-621-QD-TTg-phe-duyet-Ke-hoach-phat-trien-Tong-cong-ty-Duong-sat-Viet-Nam-giai-doan-2007-2010-vb20561t17.aspx.

Prime Minister's Office of Vietnam. (2007b). *Decree no. 189/2007/NĐ-CP promulgating functions, tasks, scope of authority and organizational structure of the Ministry of Industry and Trade of Vietnam*. Retrieved from http://www.chinhphu.vn/portal/page?_pageid=33,638900&_dad=portal&_schema=PORTAL&docid=53799.

Prime Minister's Office of Vietnam. (2008). *Decree no. 17/2008/NĐ-CP promulgating functions, tasks, scope of authority and organizational structure of the Ministry of Construction of Vietnam*. Retrieved from http://www.chinhphu.vn/portal/page?_pageid=33,638900&_dad=portal&_schema=PORTAL&docid=57598.

Prime Minister's Office of Vietnam. (2009). *Decision no. 12/2009/NĐ-CP on the management of investment and construction*. Retrieved from http://www.moc.gov.vn/site/moc/legal?v=detail-doc&id=29485.

Quang, C. (2011). The Thu Thiem tunnel, the longest river tunnel in Southeast Asia, officially opens (in Vietnamese). *Dan Tri*. Retrieved from http://dantri.com.vn/su-kien/chinh-thuc-thong-ham-thu-thiem-ham-vuot-song-lon-nhat-dna-539989.htm.

Reinhardt, J. (1993). Industrial restructuring and industrial policy in Vietnam. In M. Than & J. L. H. Tan (Eds.), *Vietnam's dilemmas and options: The challenge of economic transition in the 1990s* (pp. 71–96). Singapore: Institute of Southeast Asian Studies.

Ronnås, P., & Sjöberg, Ö. (1990, December 12-15 1989).*Đổi mới: Economic reforms and development policies in Vietnam economic reforms and development policies in Vietnam*. Paper presented at the international symposium, Hanoi.

Schwab, K. (Ed.). (2008). *The global competitiveness report 2008–2009*. Geneva: World Economic Forum.

Schwab, K. (Ed.). (2009). *The global competitiveness report 2009–2010*. Geneva: World Economic Forum.

Than, M., & Tan, J. L. H. (1993). The Vietnamese economy in transition: Introductory review. In M. Than & J. L. H. Tan (Eds.), *Vietnam's dilemmas and options: The challenge of economic transition in the 1990s* (pp. 1–21). Singapore: Institute of Southeast Asian Studies.

Thanh, N. (2011). Maritime industry of Vietnam: Far behind its potential (in Vietnamese). Laws. Retrieved from http://phapluattp.vn/20110606104313l9p0c1014/kinh-te-bien-viet-nam-chua-tuong-xung-voi-tiem-nang.htm.

Thinh, P. (2010). The most modern road tunnels in Vietnam: 8 "scandals" within 9 months (in Vietnamese). Viet Bao. Retrieved from http://vietbao.vn/Xa-hoi/Ham-duong-bo-hien-dai-nhat-VN-9-thang-8-xicangdan/75242113/157/.

Tho, H. (2008). Conclusion on the cause of the collapse of two side spans of Can Tho Stayed-cable Bridge (in Vietnamese). Viet Bao. Retrieved from http://vietbao.vn/Xa-hoi/Ket-luan-nguyen-nhan-sap-2-nhip-dan-cau-Can-Tho/65138261/157/.

Tien, N., & Huong, L. (2009, August 4). Dozens of new leaks in the Kim Lien road tunnel. *Dan Tri*. Retrieved from http://dantri.com.vn/c0/s0-341620/hang-chuc-diem-ro-ri-moi-trong-ham-duong-bo-kim-lien.htm.

Tran, V. H. (2008). *Overview of the building material sector of Vietnam (in Vietnamese)*. Ho Chi Minh City: Italian Trade Commission.

Tran, T. (2009). Sudden surge in FDI and infrastructure bottlenecks: The case in Vietnam. *ASEAN Economic Bulletin, 26*(1), 58–76.

Tuan, Q. (2010). Quality of construction works. *Nhan Dan*. Retrieved from http://www.nhandan.com.vn/cmlink/nhan-dan-i-n-t/th-i-s/chinh-tr/cung-suy-ng-m/ch-t-l-ng-cong-trinh-xay-d-ng-1.270054#9xsO47MTdKJK.

United Nations. (2010a). Selected statistics about Viet Nam. Retrieved February 16, 2011, from http://www.un.org.vn/en/about-viet-nam/basic-statistics.html.

United Nations. (2010b). Viet Nam at a glance. Retrieved November 28, 2010, from http://www.un.org.vn/en/about-viet-nam/overview.html.

United Nations Statistics Division. (2010). National accounts estimates of main aggregates. Retrieved December 15, 2010, from http://data.un.org/Browse.aspx.

Van, D. (2010a). Alarming quality of a VND 300 million road. *Dan Tri*. Retrieved from http://dantri.com.vn/c36/s20-396852/hoang-hon-voi-chat-luong-con-duong-tri-gia-hon-300-ty-dong.htm.

Van, D. (2010b). "The multi-hundreds-of-million-dong road" deteriorates: Additional asphalt to solve the issue? *Dan Tri*. Retrieved from http://dantri.com.vn/c20/s20-425867/con-duong-tram-ty-xuong-cap-bu-nhua-len-la-on-ca.htm.

Vietnam Association of Construction Contractors. (2010). About us. Retrieved March 28, 2011, from http://www.buildviet.info/vacc/p2064c2076/gioi-thieu.htm.

Vietnam Association of Construction Contractors. (2011). Construction labour: "rob Peter to pay Paul". Retrieved from http://www.vacc.com.vn/vn/Dao-Tao/lao-dong-nganh-xay-dung-giat-gau-va-vai/.

Vietnam Association of Mechanical Industry. (2010a, November 26). "Loophole" in bidding. Retrieved from http://www.vami.com.vn/Chitiettintuc/tabid/9379/ArticleID/107313/tid/9366/Default.aspx.

Vietnam Association of Mechanical Industry. (2010b). National threat: Chinese wins 90 % of major bidding packages. Retrieved from http://www.vami.com.vn/Chitiettintuc/tabid/9379/ArticleID/107314/tid/9366/Default.aspx.

Vietnam Association of Mechanical Industry. (2010c, July 28). Targets of the development plan for mechanical sector failed in the first 10 year period. Retrieved from http://www.vami.com.vn/Chitiettintuc/tabid/9379/ArticleID/107261/tid/9366/Default.aspx.

Vietnam Bridge and Road Association. (2011). About us. Retrieved March 20, 2011, from http://www.cauduongcang.com/quangcao/thongtin/72D61B.aspx.

Vietnam Business Forum. (2010). About Vietnam business forum. Retrieved December 3, 2010, from http://www.vbf.org.vn/about.aspx?x=1&c=1.

Vietnam Business Forum. (2011). How to reinforce the efficiency and the leading role of state-owned corporations and economic groups. Retrieved from http://vccinews.vn/?page=detail&folder=165&Id=3097.

Vietnam Expressway Corporation. (2009). About us. Retrieved March 30, 2011, from http://www.expressway.com.vn/.

Vietnam Federation of Civil Engineering Associations. (2011). Professional associations. Retrieved March 15, 2011, from http://www.tonghoixaydungvn.org/Default.aspx?Tab=421&Tinso=166.

Vietnam Financial Review. (2010). Vietnam's transportation infrastructure market update. *Vietnam Financial Review*. Retrieved from http://www.vfr.vn/index.php?option=com_content&task=view&id=1195&Itemid=4.

Vietnam Railways. (2012). Development history. Retrieved January 25, 2013, from http://www.vr.com.vn/thong-tin/lich-su-phat-trien.html.

Vo, T. T. (2005). Vietnam's trade liberalization and international economic integration: Evolution, problems, and challenges. *ASEAN Economic Bulletin, 22*(1), 75.

Vo, T. T. (2007). Challenges, prospects and strategies for CLMV development: The case of Vietnam. In C. Sotharith (Ed.), *Development strategy for CLMV in the age of economic integration* (pp. 497–515). Indonesia: Economic Research Institute for ASEAN and East Asia.

Vo, V. T. (2008, September 23). Infrastructure development: Unnecessary bureaucracy causes huge waste. *Tuoi Tre Online*. Retrieved from http://tuoitre.vn/Chinh-tri-Xa-hoi/279793/Dau-tu-xay-dung-co-ban-Thu-tuc-ruom-ra-gay-lang-phi-lon.html.

Vo, D. L., Dao, L. M., Nguyen, T. Q., Do, D. D., Le, V. S., Le, B. L., et al. (1995). *Vietnam's economic renovation and foreign economic policies*. Hanoi: Social Sciences Publishing House.

Vo, T. T., & Nguyen, A. D. (2009). Vietnam after two years of WTO accession: What lessons can be learnt? *ASEAN Economic Bulletin, 26*(1), 115–135.

Vu, T. (2008, August 28). Quality incident in the Thu Thiem tunnel. *Giao Thong Van Tai*. Retrieved from http://www.giaothongvantai.com.vn/Desktop.aspx/News/Dau-tu-xay-dung-ha-tang-giao-thong/Su_co_cong_trinh_ham_Thu_Thiem/.

Warlters, M. (2006). *Vietnam's infrastructure challenge—infrastructure strategy: Cross-sectoral issues* (working paper no. 37184). Washington, D.C.: World Bank.

World Bank. (2010). Data by country: Vietnam. Retrieved November 20, 2010, from http://data.worldbank.org/country/vietnam.

World Bank Data. (2013). World development indicator. Retrieved March 28, 2013 from http://data.worldbank.org/country/vietnam.

Chapter 7
Research Methodology

Abstract This chapter discusses the research methodology adopted in this study. This chapter starts with a review of the use of qualitative and quantitative research strategies within the literature of public administration. This section is followed by a brief description of the research design selected for this study. Details of the approaches used for data collection and data analysis for this study are then discussed. The focus is on the efficient use of public investments in infrastructure to support trade and economic development, as well as the influencing factors.

Keywords Research methodology · Qualitative methods · Quantitative methods · Factor analysis · Content analysis

7.1 Introduction

This chapter describes the research methodology used in this study. Since government planning is one of the issues of the broader public administration research, the chapter starts with a review of the use of qualitative and quantitative research strategies within the literature of public administration. This section is followed by a brief description of the research design selected for this study. Details of the approaches used for data collection and data analysis for this study are then discussed. The focus is on the efficient use of public investments in infrastructure to support trade and economic development, as well as the influencing factors.

7.2 Qualitative Versus Quantitative Methods

The study of public administration used multiple research methods and techniques (Adams and White 1994; Lan and Anders 2000). Generally, these methods can be divided into two major approaches, namely quantitative and qualitative (McNabb 2002). More specifically, the methods employed include the following:

G. Dang and L. Sui Pheng, *Infrastructure Investments in Developing Economies*,
DOI 10.1007/978-981-287-248-7_7

- Quantitative methods: descriptive statistics and/or simple correlation analysis; intermediate statistics (hypothesis testing and some inferential statistics); and more advanced statistical tests including regression analysis, time series and more sophisticated inferential statistics;
- Qualitative methods: case study; ethnography; and others (e.g. literature reviews and reports of interviews) (Lan and Anders 2000).

There has been a quantitative versus qualitative debate. Scientists, who were in favour of quantitative methods, argued that an idea or concept was not meaningful if it was unable to be measured or expressed in numbers. Meanwhile, qualitative methods owed their growth in popularity in researches required to uncover complex cause–effect relationships, processes and structures in their natural settings as experienced by the actors themselves (Tan 2007). A survey by Lan and Anders (2000), however, showed that the frequency of studies using quantitative method and that of studies using qualitative method were comparable. Since no single method has proved to be more useful than others within the literature of public administrative research, Lan and Anders (2000) argued that the major issue concerning research methodology should be about how to use a method appropriately rather than on which methods should be used. In order to use a method appropriately, it is important for researchers to consider all relevant characteristics as well as strengths and weaknesses of these methods.

Quantitative and qualitative research methodologies are mainly distinguished by the ways of describing and processing research data. Quantitative methodologies often involve the use of statistical analysis of data to answer research questions, which is often to measure and analyse causal relationship between variables. This approach thus requires a sample size much larger than the one used by the qualitative approach to ensure that statistically, the sample can be a good representative [Carey (1993) as cited by Sale et al. (2002)]. Techniques that quantitative methodologies usually utilize to collect data include experimental design and administered survey with a limited range of predetermined responses (McNabb 2002). Therefore, the advantage of this approach is that its standardized data collection task is less time-consuming and useful to study a large sample size, thus facilitating comparison and statistical aggregation of the data, as well as making generalized research findings possible. This approach is consequently useful in studies focusing on phenomenon predicting or explaining (Johnson and Onwuegbuzie 2004).

Qualitative methodologies focus on interpretation and understanding rather than predicting and explaining (Miller and Whicker 1999). Two most often used methods in the social and administrative sciences are ethnography and case study (McNabb 2002). The approach involves the use of in-depth and focus-group interviews, participant observation and analysis of documents and/or artefacts in order to collect data (McNabb 2002). As a result, qualitative method relies on the texts and discourses of participants involved. Samples used by this approach are small and purposeful samples that are capable of delivering important information. Samples are thus not supposed to represent large populations, but to represent detailed snapshots of the samples under study [Reid (1996) as cited by Sale et al.

(2002)]. Major advantages of this approach are that it is useful for describing complex phenomena, thus providing individual case information with rich details. However, the method is more time-consuming, and the knowledge obtained may not be generalizable to other case (Johnson and Onwuegbuzie 2004).

Recent studies have argued that both methods are important and could be used in a single research study (Sale et al. 2002; Johnson and Onwuegbuzie 2004). A combined research strategy, a mixture of both qualitative and quantitative methods and techniques is possible. The goal of combining these two methods is not to replace either of these approaches but rather to strengthen the advantages and minimize the disadvantages of both. Each method is used as a way to expanding one's understanding of a phenomenon. Generally, a mixed-method strategy is selected for complementary purposes that a single method cannot accomplish. Combining research methods is thus useful in some areas of research where the complexity of phenomena requires data from a large number of perspectives. In addition, the generalizability of the results can increase using this approach. However, a mixed-method strategy requires more resources than a single-method strategy and is therefore more expensive, more time-consuming and requiring more human resources. It is also more difficult for researchers to learn about multiple methods and approaches and understand how to mix these appropriately (Johnson and Onwuegbuzie 2004).

7.3 Research Design

Selection of research design mainly depends on the research objectives and questions. The approach selected is the one that provides the most needed data to answer research questions and ensures that the objectives of the study are achieved (Crabtree and Miller 1999; Sale et al. 2002; Yin 2009). Surveys, in-depth interviews and case studies are common in social science and management spheres (Fellows and Liu 2003). Surveys are quick and efficient to explore particular issues (Tan 2007). Meanwhile, in-depth interviews and case studies are essentially to examine why and how a decision or a set of decisions were taken, and implemented, and with what result (Schramm 1971; Morse 1994). Studies focusing on infrastructure planning usually employ the mixed use of data (e.g. documents, interviews and observations) to provide a stronger basis for analysis and interpretation of related government policies (see e.g.: Hossain 2000; Webster and Theeratham 2004; Mustajab 2009).

Following the traditional approach, a questionnaire survey was adopted in this current research to capture the views of the concerned parties on the research issues. In-depth interviews involved case studies, and document analyses were also employed in this research. In addition, a longitudinal survey was used to study the efficiency of public investments in infrastructure over time. The longitudinal survey was conducted before the survey and interviews were carried out. The purpose of the longitudinal survey is to provide an overview of the developmental trends of the use of funded infrastructure construction to support economic development in

Vietnam. Notably, the longitudinal survey does not attempt to establish the linkages between the use and potential "causal" factors, but aims to explore the extent to which public investments in infrastructure are efficiently used in Vietnam.

A semi-structured questionnaire was used to collect the data in this research. A semi-structured questionnaire is designed with open-ended questions, thus not limiting the respondents' choice of answers (McCracken 1988; Gubrium and Holstein 2002). The disadvantages of this type of questionnaire are that responses are time-consuming, coding is difficult, and more effort from the respondents is required (Oppenheim 1992). However, compared to structured and unstructured questionnaires, semi-structured questionnaires fit the study's needs with questions that help and direct the respondent into the research topic area and provide the opportunity for the researcher to gather more information in depth whenever appropriate (McCracken 1988; Creswell 1998).

In-depth interviews were conducted to examine how to resolve the issues found from the survey. The population of interest for in-depth interviews is therefore key planners, policy-makers and key players from the non-governmental sector. Besides, in-depth interviews involved some relevant case studies and document analyses.

7.4 Longitudinal Survey

As mentioned earlier, the purpose of the longitudinal survey is to provide an overview of the development trends of the use of public investments in infrastructure to support trade and economic growth. The survey was carried out by examining the efficiency of public investments in infrastructure construction to support trade and economic growth over time.

Public sector expenditure efficiency analysis is used to compare the relative outcomes (effects) and costs of the relevant public expenditure. Public sector expenditure efficiency is expressed in terms of a ratio where the numerator is a performance level (a gain in performance from a measure, e.g. the competitiveness index), and the denominator is the amount of public expenditure used to achieve a given performance level (Afonso et al. 2005). The method has been widely used for efficiency analysis of public spending such as those completed by Pang and Herrera (2005), Mandl et al. (2008) and Borge et al. (2008). Since this study focuses on efficiency analysis of public investments in infrastructure construction to support trade and economic growth, the method is particularly relevant.

From the literature review, the objective of public investments in infrastructure to support trade is to improve the quality and the extensiveness of infrastructure networks for trading expansion, subsequently economic growth. Public investments in infrastructure construction expressed as a share of GDP can be assumed to reflect costs of achieving the objective. Therefore, in order to examine the efficiency of public investments in infrastructure construction to support trade and economic growth, the ability of infrastructure to support trade and economic growth was weighted by the amount of public investments in infrastructure construction.

International organizations (e.g. International Road Federation, International Air Transport Association and International Telecommunication Union) often use a set of indicators for assessing infrastructure development for a given means of transport, including passenger-carrying capacity, freight-carrying capacity, infrastructure density and level of infrastructure investments. While these indicators reflect the level of infrastructure development for a given means of transport in many ways, it is not useful when a single and integrated index for assessing infrastructure development as a whole is needed. Such an integrated index, however, can be found in the Global Competitiveness Report published annually by the World Economic Forum (WEF). Besides separate indices for different types of infrastructure, the WEF uses a composite index for assessing the competitiveness of a nation in terms of infrastructure development. The indicator reflects the quality and extensiveness of infrastructure networks that integrate the national market and connecting it at low cost to markets in other countries and regions (Schwab 2009). The indicator offered by the WEF is thus particularly useful in this study. Moreover, data collected from WEF's Global Competitiveness Report are more reliable as these were backed by eminent Harvard academics Professor Jeffrey Sachs and Professor Michael Porter. Data quality in the Global Competitiveness Report has improved over the years with stronger academic foundation and methodology (Schwab 2008). Data from the Global Competitiveness Report continue to be used as a significant source of policy discussion about related issues in developing countries (Dollar et al. 2002; Wilson et al. 2003, 2006; Arvis et al. 2007). As a result, the competitiveness index of Vietnam in terms of infrastructure development as reported by the WEF was used as a proxy for the ability of infrastructure to support trade and economic growth in this study.

Data of the competitiveness index of Vietnam in terms of infrastructure development were thus collected from the Global Competitiveness Report. Data of public investments in infrastructure construction were collected from the General Statistics Office of Vietnam (GSO) and the Ministry of Planning and Investment of Vietnam, as well as other official documents, related studies, reports and articles. With the data collected, the efficiency of public investments in infrastructure construction to support trade and economic growth was computed on a yearly basis (see Sect. 8.2). Microsoft Excel software was utilized to assist in the data analysis.

7.5 Questionnaire Survey

7.5.1 Data Collection

The survey was divided into two parts. The first part is used to examine the perceptions of achievements of public investments in infrastructure for supporting trade expansion, and subsequently, economic growth among the actors and stakeholders involved. From the literature review, government's investments in infrastructure construction play an important role in providing quality and extensiveness of infrastructure networks that integrate the national market and connecting it at low cost to markets in other countries and regions. However, the results found in the majority of infrastructure projects were cost overruns, benefit shortfalls and waste (Flyvbjerg 2007). As reviewed in Chap. 5, the outcomes of the inefficient use of public investments in infrastructure in the planning and implementation of infrastructure development can be described as poorly selected and poorly executed infrastructure investments and ultimately waste of national resources. As a result, to elicit the perceptions of the achievements, respondents were asked to rate against a five-point Likert scale, from "not significant" to "extremely significant" about the occurrence of these unexpected outcomes.

The second part of the survey was conducted to explore the factors affecting the use of public investments in infrastructure for supporting economic development. Derived from the literature review, these factors are classified into the following major groups:

- Infrastructure planning:
  - Capacity for estimating and monitoring of rates of return of infrastructure projects,
  - Politicized decision-making,
  - Transparency and accountability problems,
  - Institutional weaknesses in planning.
- Infrastructure delivery:
  - Political commitment,
  - Corruption issues in infrastructure construction,
  - Land acquisition problems,
  - Building capacity of local firms,
  - Institutional weaknesses in infrastructure construction.

Similarly, the five-point Likert scale was applied for questions relating to the degree of occurrence and influence of factors. These questions were then followed by open-ended questions asking the respondents about further issues based on their experience.

A sample of these questions is attached in Appendix A. The cover letter that provides a brief description of the survey and its goals is also included. A pretest was carried out before the actual survey was conducted. A small sample of respondents from the population of interest was randomly selected. The pretest respondents included two senior government officials and two seniors working in infrastructure construction-related firms headquartered in Hanoi. Besides, two professors working for research institutions in Hanoi joined the pretest to enhance the validity of the research. The pretest respondents were requested to give their feedbacks on the relevance of questions. In general, all pretest respondents agreed that the questionnaire was comprehensible. However, they were concerned that requiring respondents' particulars, especially respondents' names, might make them reluctant to answer the questionnaires. That is because the issue of the efficiency of government funding invested in infrastructure is quite sensitive. Many Vietnamese, especially those who work for the government, may not feel comfortable to take part in the survey. Other concerns were related to phrasing. The unexpected outcomes mentioned in Part I should be more specific. The expression “Political discontinuity” in Part III could make respondents confused with changes of leading party and government. Since there is no change of leading party in Vietnam since independence, it should be better to replace the expression with “changes in the political affiliations of the government responsible for the projects”. Based on these suggestions, the questionnaire was modified before the actual launch of the survey.

The purpose of the survey is to examine perceptions of issues and challenges of the current use of public investments in infrastructure among the actors involved. These actors can be classified into two groups: governmental and non-governmental actors. The population of interest thus consists of both governmental and non-governmental actors.

Since the level that this study focuses on is national level, the governmental actor group thus consists of planners and policy-makers working in central government agencies, which have a role in policy-making as well as planning for construction, infrastructure and international trade of Vietnam as highlighted in Chap. 6, including relevant departments and agencies under the Ministry of Planning and Investment, Ministry of Construction, Ministry of Transport, Ministry of Information and Communications, and Ministry of Industry and Trade as well as the ministries’ think tanks. The initial list of agencies was discussed with policy experts from the central government to see whether there was any omission (see Table 7.1).

In total, there are 261 units and divisions under relevant departments of the above-mentioned ministries (see Table 7.1). The list of these divisions and departments can be found in various Websites of the related ministries. The list served as the sampling frame for this study. Based on the list, a few respondents from these government agencies were contacted and asked to provide referrals for additional respondents. Selected individuals were planners and policy-makers with at least 5 years of work experience since they would be senior enough to provide both a technical and a policy perspective.

In addition, this study is interested in perceptions of non-governmental actors who are involved in the implementation of government-funded infrastructure projects. The non-governmental actor group in this study thus refers to the actors other than the planners and policy-makers, including financial institutions, traders, developers, consultants and contractors. There are 178 banks, financial institutions, and equipment rental companies. The list of these organizations (with contacts), which served as the sampling frame for financial institutions, is posted on the Website of the State Bank of Vietnam (State Bank of Vietnam 2011). The list of the most prominent traders in Vietnam is similarly posted on the Website of the Ministry of Industry and Trade. The list of the 217 traders (with contact details) served as the sampling frame for traders (Ministry of Industry and Trade 2011).

According to a report by the Ministry of Construction, there are 55,870 firms in the construction sector (Hoang 2013). The sampling frame for local developers, consultants and contractors is the list of major developers, consultants and contractors in the infrastructure sector posted on the Websites of the Ministry of

Table 7.1 Number of divisions and units under relevant ministerial departments

Departments	Units
Under the MPI	16
Department for National Economic Issues	
Department of Infrastructure and Urban Centers	
Department of Investment Supervision and Appraisal	
Department of Planning Management	
Public Procurement Agency	
Under the MoC	29
Department of Architecture and Spatial Planning	
Urban Development Agency	
Infrastructure Facility Agency	
State Authority for Construction Quality Inspection	
Department of Construction Economics	
Department of Science, Technology and Environment	
Department of Building Materials	
Department of Finance and Planning	
Department of Legal Affairs	
Department of Investment and Construction Management	
Department of Housing and Real Estate	
Under the MoT	101
Department of Planning and Investment	
Directorate for Roads of Vietnam	
Vietnam Railway Administration	
Vietnam Inland Waterways Administration	
Vietnam Maritime Administration	
Civil Aviation Administration of Vietnam	
Transport Engineering Construction and Quality Management Bureau	
Under the MoNRE	14
General Directorate of Land Registry	
Under the MoIC	16
Department of Telecommunications	
Under the MoIT	85
Department of International Cooperation	
Multilateral Trade Policy Department	
Vietnam Trade Promotion Agency	
Department of Export and Import	
Vietnam Competition Authority	
Vietnam e-Commerce and Information Technology Agency	
Market Control Department	
Asia-Pacific Market Unit	
Europe Market Unit	
America Market Unit	
Africa, West Asia and South Asia Market Unit	
Total	261

Source Ministry of Construction (2010), Ministry of Industry and Trade (2010), Ministry of Information and Communications (2010), Ministry of Natural Resources and Environment (2010), Ministry of Planning and Investment (2010), Ministry of Transport (2010)

Construction, the Ministry of Transport and the Ministry of Information and Communication. There are 19 major developers and 326 consultants and contractors in the list (Ministry of Information and Communications 2011; Ministry of Transport 2011). The sampling process started with the original list of relevant individuals in the lists of organizations mentioned above. Each potential survey participant was personally contacted to see whether they were interested, and the contacted individuals would also be asked to suggest other individuals who were of interest to this study. Similar to respondents from government agencies, respondents from firms with at least 5 years of work experience were selected.

The population of the survey includes professionals in government agencies as well as firms engaged in large-scale infrastructure projects. Since all major government agencies are headquartered in Hanoi, most questionnaires were delivered to professionals in Hanoi. Questionnaires can be delivered via mail, telephone or directly delivered. Mail and telephone delivery can save time and cost, while personal delivery can increase response rate (Alston and Bowles 1998; Jackson 2006). For this study, in order to increase response rate and ensure the accuracy of answers, questionnaires were directly delivered to professionals in Hanoi. Only a small number of questionnaires were delivered to professionals in Ho-Chi-Minh City via email to save cost and time.

7.5.2 Data Analysis

The simple data analysis of responses was first conducted by ranking the problems in terms of degree of occurrence and level of influence. The analysis also examined whether there was any difference in perceptions of different groups of respondents by using rank correlation. Scores, therefore, were first converted to ranks. Based on the ranks, the Spearman's rank correlation coefficient would be computed to test the strength of associations between the rankings of respondent groups using the following formula (Tan 2007):

$$r_s = 1 - \frac{6\sum d_i^2}{n^3 - n}$$

where d_i is the difference between ranks for *i*th characteristic and n is the number of characteristics ranked. For large samples ($n > 20$), a test of significance for r (with H_o: the correlation is not significant) is given by Tan (2007)

$$z = r_s\sqrt{(n-1)} \sim N(0,1)$$

Further data analysis of this study involved the analysis of 35 variables in order to identify key factors affecting the efficiency of government-funded infrastructure construction. In order to identify key factors, the degree of criticality was considered first. By considering the degree of criticality first, the study could eliminate factors that are not perceived as critical. Those factors that are perceived as critical would then be selected for further data analysis of their degree of occurrence. Data analysis is thus first performed to extract the problems that are perceived as critical by the respondents based on their degree of criticality. In the next step, the relationships between the occurrences of the key problems and the outcome or effect are examined. Since the data analysis involved simultaneous analysis of the large number of variables, the application of multivariate statistical techniques is considered. The selection of an appropriate multivariate technique depends on how variables are classified as shown by Hair et al. (2010). Without a reason to divide the variables in this study into dependent and independent classifications, an interdependence technique should be utilized in factor analysis.

Among interdependence techniques, factor analysis and structural equation modelling (SEM) are useful for studies focusing on variables rather than on cases or respondents. Factor analysis is more appropriate in analysing the structure of variables by reducing a large number of variables into a smaller set of factors. Meanwhile, SEM is more appropriate in analysing the cause–effect relations between factors (Hair et al. 2010). As a result, in the first stage, factor analysis was used in this study to extract key causal factors limiting the efficient use of public investments in infrastructure. In the second stage, SEM was used to examine the relationships between these key causal factors and the outcomes; the dependent variable is a measure of the occurrence of the outcomes of the inefficient use of public investments in infrastructure, and the independent variables are measures of the occurrence of the factors. Measures of the outcomes and factors are thus the respondents' ratings of the occurrence of the outcomes and factors.

Using factor analysis to extract key factors from a large number of variables with p variables and k common factors to be determined and a unique factor δ, the factor analytic model can be written as follows:

$$
\begin{aligned}
X_1 &= l_{11}F_1 + l_{12}F_2 + \cdots + l_{1k}F_k + \delta_1 \\
X_2 &= l_{21}F_1 + l_{22}F_2 + \cdots + l_{2k}F_k + \delta_2 \\
&\ldots \\
X_p &= l_{p1}F_1 + l_{p2}F_2 + \cdots + l_{pk}F_k + \delta_p
\end{aligned}
$$

The coefficients (ls) reflect the contribution of each factor to each variable, thus called factor loadings. In matrix form,

$$\mathbf{x} = \mathbf{Lf} + \boldsymbol{\delta}$$

where $\mathbf{x}$ is a $p \times 1$ vector, $\mathbf{R}$ is the correlation matrix, $\mathbf{x} \sim \mathbf{N}(\mathbf{0}, \mathbf{R})$, and $\mathrm{Var}(x_i) = 1$; $\mathbf{L}$ is a $p \times g$ matrix of factor loadings; $\mathbf{f}$ is a $g \times 1$ vector of independent common factor and $\mathbf{f} \sim (\mathbf{0}, \mathbf{I})$; $\boldsymbol{\delta}$ is a $p \times 1$ vector of unique factors, $\boldsymbol{\delta} \sim (\mathbf{0}, \boldsymbol{\Psi})$, or $\mathrm{Var}(\delta_i) = \psi_i$; and $\mathbf{f}$ and $\boldsymbol{\delta}$ are independent. Hence,

$$\begin{aligned}\mathrm{Var}(x_i) &= \mathrm{Var}(l_{i1}F_1 + l_{i2}F_2 + \cdots + l_{ig}F_g + \delta_i)\\ &= l_{i1}^2\mathrm{Var}(F_1) + \cdots + l_{ig}^2\mathrm{Var}(F_g) + \mathrm{Var}(\delta_i)\\ &= \sum l_{ij}^2 + \psi_i = 1\end{aligned}$$

In the matrix form proposed by Tan (2007),

$$\begin{aligned}\mathrm{Var}(x) = R = E[xx'] &= E\left[(Lf + \delta)(Lf + \delta)'\right]\\ &= E[(Lf + \delta)(f'L' + \delta')]\\ &= E[Lff'L' + \delta\delta']\\ &= LL' + \Psi\end{aligned}$$

This relation is used to estimate factor loadings. The first step is to compute R:

$$R = \lambda_1 u_1 u_1' + \cdots + \lambda_p u_p u_p'$$

where λs are eigenvalues of R and us are corresponding eigenvectors. Eigenvalues greater than 1 are used to identify the number of factors (g) to retain. Then,

$$LL' = \lambda_1 u_1 u_1' + \cdots + \lambda_g u_g u_g'$$

and $\psi = R - LL'$ (Tan 2007).

Principle component analysis is used to extract factors, and VARIMAX rotation is used to rotate factor loadings so that they are close to 0 and 1 to facilitate factor interpretation (Hair et al. 2010). With the large number of variables and data, the use of software programs is considered. For statistical analysis, the Statistical Package for Social Sciences (SPSS) is chosen for factor analysis in this study. There are some common statistical packages, such as Minitab, SPSS, and SYSTAT. However, the SPSS is a comprehensive statistical package widely used by social scientists (Tan 2007). By utilizing this software package, it is possible for researchers to:

- import data from a variety of formats;
- retrieve and view a pre-existing data file;
- manipulate and perform statistical analysis; and
- produce joint configurations of tables and graphs (Bryman and Cramer 2009).

Utilizing this software, raw data were first entered into a computer and frequencies and percentages were calculated. Tables and cross-tabulations were then constructed in order to examine the relationship between variables.

The next stage involved the SEM to identify the relationships between the occurrences of key factors extracted from the factor analysis stage and the outcomes of the inefficient use of public investments in infrastructure. Causal factors that were statistically perceived as significant in the factor analysis were used as the independent variables in SEM. The outcomes of the inefficient use of public investments in infrastructure were the dependent variables in SEM. A single regression analysis can only interpret a single relationship between two variables. However, in a model with a number of interrelated variables, it would not be meaningful for the relationships to be interpreted separately. SEM is thus more useful to simultaneously examine the structural relationships. SEM examines the structure of interrelationships expressed in a series of equations, which depict all of the relationships between constructs (the dependent and independent variables) involved in the analysis (Hair et al. 2010). As multiple relationships are expected between variables in this study, SEM is useful to indicate which relationships are critical among variables and which require more attention.

There are two distinct approaches to carry out SEM analysis: covariance-based SEM (CB-SEM) and partial least squares SEM (PLS-SEM). CB-SEM technique estimates model parameters by minimizing the difference between estimated covariance values and observed covariance values (Hair et al. 2010). In order to apply CB-SEM, several assumptions of data and minimum sample size need to be met. Compared to CB-SEM, PLS-SEM has less stringent assumptions of data and minimum sample size. Statistically, PLS-SEM also uses different approaches to produce parameter estimates. PLS-SEM estimates model parameters by maximizing the explained variance of the dependent latent constructs. While CB-SEM has been widely applied in marketing researches, the number of studies using PLS-SEM is growing. On the one hand, CB-SEM is appropriate if the research objective is about theory confirmation. On the other hand, PLS-SEM proves more useful if the research objective is about prediction or more about an extension of an existing structural theory. Moreover, when data characteristics or sample size do not meet the CB-SEM assumptions, PLS-SEM estimates can be used as proxies of CB-SEM results. Another advantage of PLS-SEM is the ability to produce estimates of factor scores which are specifically valuable in impact-performance analyses. Based on factor scores, key factors can be identified for potential performance improvements (Hair et al. 2011). Given the research objective is to uncover key issues affecting the investment efficiency of funded infrastructure development, PLS-SEM is more appropriate for this study. Among software packages executing PLS-SEM, Smart-PLS software can conduct all data analyses required in PLS-SEM (Hair et al. 2011). The software is thus selected in this study. With a path model example, the stages and steps in calculating the basic PLS-SEM algorithm are presented in Fig. 7.1.

The path model example shown in Fig. 7.1 has two components. The structural model (the inner model) in the PLS-SEM context refers to the relationships between latent variables y_g (LVs). The measurement model (the outer model) refers to the relationships between latent variables and observed variables ($x_1, \ldots, x_3$). The following PLS-SEM algorithm is adopted from Monecke and Leisch (2012).

In the measurement model, the latent variables y_g (LVs) are measured by the block $\boldsymbol{X}_g$ consisting of three observed variables such as x_1, x_2 and x_3 (MVs) in the model. Assuming that all MVs contained in the data matrix $\boldsymbol{X}$ are scaled to have zero mean and unit variance, and each block of MVs $\boldsymbol{X}_g$ is already transformed to be positively correlated for all LVs y_g, $g = 1, \ldots, G$. Each block of MVs reflects its LV and can be written as the multivariate regression:

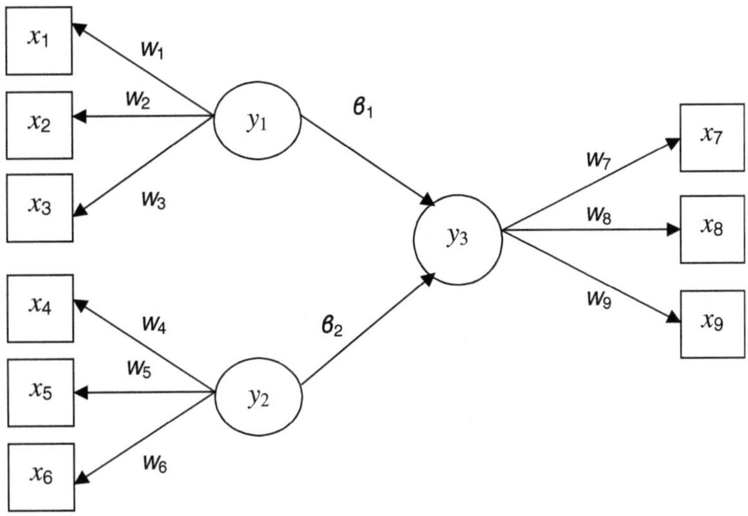

Fig. 7.1 Path model example

$$X_g = y_g w_g^\top + F_g, \quad E[F_g | y_g] = 0$$

So outer weight w_g^T can be estimated by least squares as

$$\begin{aligned}\widehat{w}_g^\top = \left(y_g^\top y_g\right)^{-1} y_g^\top X_g &= \mathrm{VAR}(y_g)^{-1}\mathrm{COV}(y_g, X_g) \\ &= \mathrm{COV}(y_g, X_g) \\ &= \mathrm{COR}(y_g, X_g).\end{aligned}$$

Assigning $\kappa_g = \{k \in \{1, \ldots, K\} | x_k \sim y_g\}$ to be a set of indices for MVs related to LV y_g, the w_g, g = 1, …, g, is a column vector of length $|\kappa_g|$. Adjacency matrix $\boldsymbol{M}$ for the measurement model has the same structure as the matrix of outer weights $\boldsymbol{W}$ and is used for the initialization. With m_{kg} = 1, MV x_k is one of the indicators of LV. Calculating the basic PLS-SEM algorithm thus starts with matrix $\boldsymbol{M}$ used for the initialization:

- Step 1: Initialization (constructing each LV as a weighted sum of their MVs): scaling LVs to have unit variance.

$$\hat{Y} = XM$$

$$\hat{y}_g = \frac{\hat{y}_g}{\sqrt{\mathrm{VAR}(\hat{y}_g)}}, \qquad g = 1, \ldots, G$$

- Step 2: Inner approximation $\tilde{Y}$ (estimating each LV as a weighted sum of its neighbouring LVs) as matrix product of the outer estimate $\hat{Y}$ and the matrix of inner weights E: again scaling computed LVs to have unit variance.

$$\tilde{Y} = \hat{Y}E$$
$$\tilde{y}_g = \frac{\tilde{y}_g}{\sqrt{\mathrm{VAR}(\tilde{y}_g)}}, \qquad g = 1, \ldots, G$$

- Step 3: Outer approximation [recalculating the weights on the basis of the LV values from the inner approximation (Step 2)].

$$\begin{aligned}\widehat{w}_g^{\top} &= \left(\tilde{y}_g^{\top}\tilde{y}_g\right)^{-1}\tilde{y}_g^{\top}X_g \\ &= \mathrm{COR}\left(\tilde{y}_g, X_g\right).\end{aligned}$$

- Step 4: Outer estimation (estimating the factor scores by means of the MVs)

$$\widehat{Y} = XM$$
$$\widehat{Y}_g = \frac{\widehat{Y}_g}{\sqrt{\mathrm{VAR}\left(\widehat{Y}_g\right)}} \qquad g = 1, \ldots, G$$

- Step 5: If the relative change of all the outer weights from one iteration to the next is smaller than a predefined tolerance,

$$\left|\frac{\widehat{w}_{kg}^{\mathrm{old}} - \widehat{w}_{kg}^{\mathrm{new}}}{\widehat{w}_{kg}^{\mathrm{new}}}\right| < \mathrm{tolerance} \quad \forall k = 1, \ldots, K \quad \wedge g = 1, \ldots, G$$

  The estimation of factor scores done in Step 4 is taken to be final. Otherwise, go back to Step 2.
- Step 6: Estimates of path coefficients (β_g). The estimates can be done by ordinary least squares (OLS). For each LV$\hat{y}_g$, $g = 1, \ldots, G$, the path coefficient is the regression coefficient on its predecessor set $\widehat{y}_g^{\mathrm{pred}}$.

$$\begin{aligned}\hat{\beta}_g &= \left(\hat{y}_g^{\mathrm{pred}\top}\hat{y}_g^{\mathrm{pred}}\right)^{-1}\hat{y}_g^{\mathrm{pred}\top}\hat{y}_g \\ &= \mathrm{COR}\left(\hat{y}_g^{\mathrm{pred}}, \hat{y}_g^{\mathrm{pred}}\right)^{-1}\mathrm{COR}\left(\hat{y}_g^{\mathrm{pred}}, \hat{y}_g\right)\end{aligned}$$

The estimated matrix of path coefficients $\widehat{B}$

$$\widehat{\beta}_{ij} = \begin{cases} \widehat{\beta}_{gj}, & \text{for } j \in y_i^{\text{pred}}, \\ 0\ , & \text{else.} \end{cases}$$

The matrix of total effects $\widehat{T}$ as the sum of the 1-to-G-step transition matrices:

$$\widehat{T} = \sum_{g=1}^{G} \widehat{B}^g$$

The cross- and outer loadings are estimated as (Monecke and Leisch 2012):

$$\widehat{\Lambda}^{\text{cross}} = \text{COR}\left(X, \widehat{Y}\right)$$

$$\widehat{\lambda}_{kg}^{\text{outer}} = \begin{cases} \widehat{\lambda}_{kg}^{\text{cross}}, & \text{if } m_{kg} = 1, \\ 0, & \text{else.} \end{cases}$$

7.6 In-depth Interviews

For in-depth interviews, open-ended questions were used to collect the data. The interviews started by asking respondents about the respondent's experience of major issues found from the questionnaire survey. Subsequently, whenever appropriate respondents were asked how these issues affected the use of public investments in infrastructure and how they should be resolved. The interviews were recorded and fully transcribed to help the researcher to recall the narrative responses for review and discussion thereafter. Original tape recordings are kept for checking original meanings and contexts whenever necessary. A sample of the interview questions is attached in Appendix B.

Determining sample size for interviews is difficult. A number of factors can determine sample sizes in qualitative studies, such as the heterogeneity of the population, selection criteria, the scope of study and the quality of the data (Morse 2000; Ritchie et al. 2003). There are no specific rules to be followed for qualitative inquiry. Generally, the concept of saturation is used as a guiding principle in determining sample sizes in qualitative studies. A study is considered to be reaching saturation when the collection of new data does not necessarily add any further insights to the overall subject being studied (Glaser and Strauss 1967). However, qualitative studies are rarely found to be able to prove how this point is achieved (Morse 1995; Bowen 2008; Mason 2010). Little guidance for estimating actual sample sizes is found (Mason 2010). For phenomenologies, Morse (1994) recommended a minimum sample of six. Creswell (1998) suggested a sample size from 5 to 25. Guest et al. (2006) also indicated that a minimum sample size of six would be sufficient for key themes to be present. Since in-depth interviews in this study mostly involved describing with depth and in detail the current challenges and response approaches from respondents' experiences, the minimum sample size that the research targeted at is six.

After the survey, in-depth interviews with key "actors" were employed. The selected respondents were those who had/have been holding key positions in the organizations involved in planning and executing national infrastructure development plans. The sampling frame was thus the same as the questionnaire survey (see Sect. 7.5.1 for sampling method). Snowball method was used to identify the respondents. Moreover, project-related documents including related reports, evaluations and other documentation would be collected as alternative sources for cross-checking.

Data analysis for open-ended questions is challenging. One of the most difficult issues in analysing open-ended survey responses is to choose an appropriate methodology. Broadly, two methodological approaches can be found in the literature on methods for analysing free-flowing text: word-based analysis methods and code-based analysis method (Ryan and Bernard 2000). Word-based analysis methods, such as keywords in context, word counts, semantic networks and cognitive maps, use words as units of analysis. These methods can be used to capture relationships between concepts by identifying co-occurrences of words or similarities in responses between individuals. They can be useful in analysing both dense and sparse types of text (Carley 1993; Stone 1997; Jackson and Trochim 2002). On the other hand, code-based analysis methods, such as grounded theory, traditional content analysis and schema analysis, use codes as units of analysis. These methods are often used in denser types of text. They are useful for organizing text data into manageable categories and making inference about a sample (Krippendorff 1980; Weber 1990). Since the study focuses more on making inference than on capturing relationships between concepts, the code-based analysis method is more suitable for this study.

Among code-based approaches, thematic content analysis is typically used in analysing verbal materials, although it can be used for non-verbal materials as well. The technique is used to extract desired information from large body of verbal materials by categorical coding schemes (Weber 1990; Burnard 1991). Although content analysis is mainly criticized due to its researcher-driven categorical coding schemes (Geer 1988; Weber 1990; Seidel and Kelle 1995), the coding schemes in this research are developed from the survey. Thematic content analysis is used in this study to extract desired information from in-depth interviews to illustrate the various issues identified from the survey.

7.7 Two-Stage Validation

The first stage of validation involved quantitative validation via statistical tests applied to the research hypothesis (see Sects. 8.3 and 8.4). Various statistical tests were carried out throughout the data analysis, including the tests of significance of

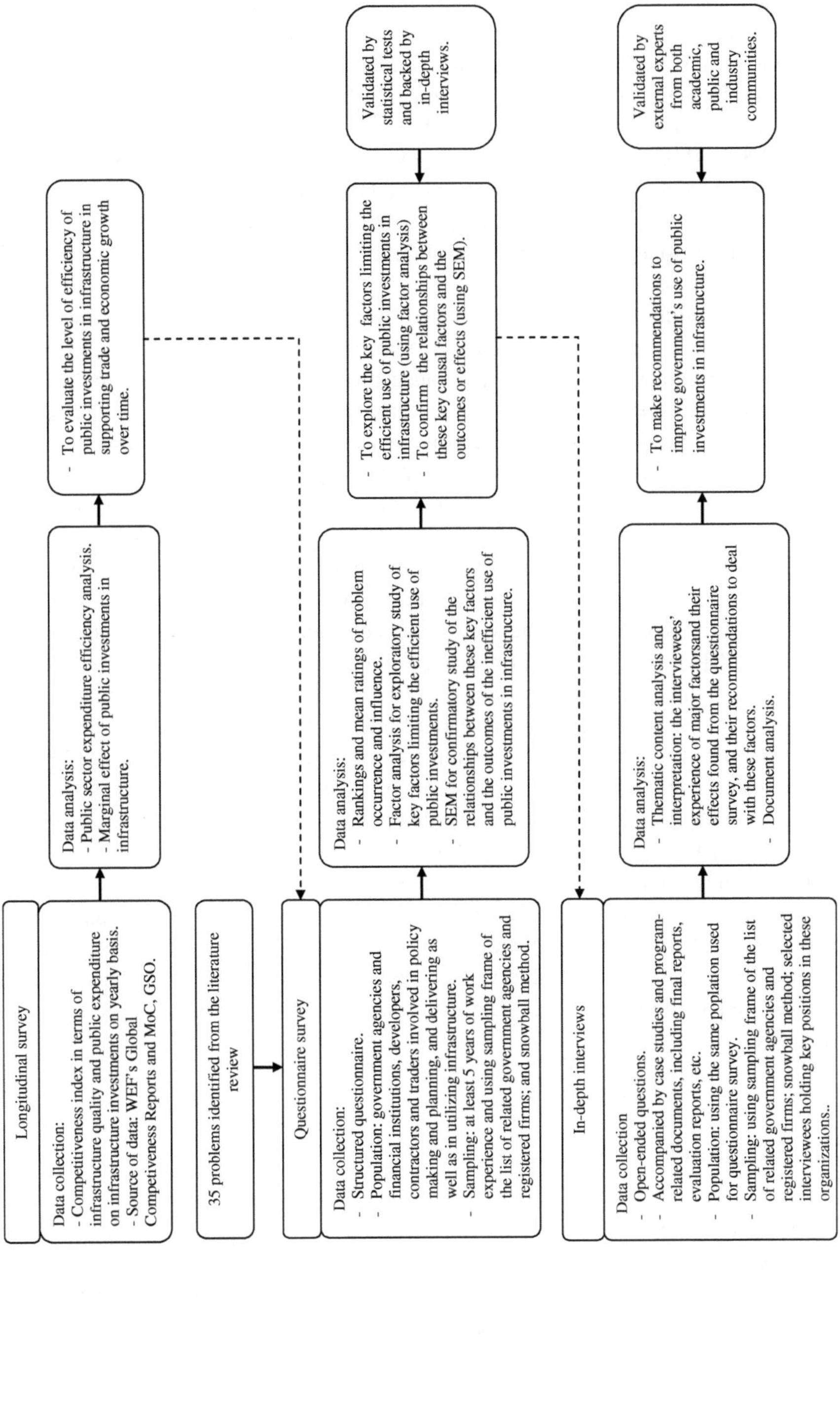

Fig. 7.2 Research design

the correlation between groups of factors, the hypothesis test on the median value of criticality of problems, the tests of the reliability and validity of the measurement model and the structural model. These tests are explained in Sects. 8.3 and 8.4.

In order to further enhance the validity of the research findings and recommendations, an independent check was undertaken at the end of the study (see Sect. 9.4). Besides the researcher, the research findings can generally be validated by research respondents or by people who are external to the study (Creswell and Miller 2000). Respondent validation is relevant when the purpose of validation is to clarify perceptions of research participants. Using the method, research findings would be returned to research participants for their feedbacks. The results of the method depend to a great extent on the stability of respondents' situation and views (Long and Johnson 2000).

Research findings can also be validated by reviewers, colleagues and interested users, who are external to the study. Among these groups, interested groups are used with particular focus on the relevance of research findings and implications (Long and Johnson 2000). Since the research findings and recommendations are relevant to the academic, public and private sectors, validation by these interested users is suitable. More specifically, the validation process was conducted through interviewing these concerned parties, including senior stakeholders from both the academic and public sectors and industry communities. Notably, selected interviewees were not research participants. The number of interviewees to validate research findings varies from study to study. For instance, the study by Mustajab (2009) was backed by 22 interviews. Seven interviewees were mentioned by Marshall (2011). Imriyas (2007) contacted five interviewees to verify the research outcomes. In the study by Hossain (2000), three interviews were conducted to ensure the internal validity of the research findings. In this research, five interviews were conducted to validate the research findings and recommendations. They are senior stakeholders with about 30–40 years of working experience. They have been engaging in the process of public investments in infrastructure since the mid-1990s. One of them was involved in the public sector, two in academics and two in the industry. They have been working in universities, research institutes, government agencies at ministerial level and the private sector. The highest educational qualification all of them hold is a doctorate. The interviews started with an explanation of the research findings. The interviewees were then asked about their feedback on the research recommendations. The discussion of their feedback is presented in Sect. 9.4.

7.8 Summary

The selection of research methodology is discussed in this chapter. The methods and approaches used in this study adopt both quantitative and qualitative analyses. The type of research design, method of sampling, data collection and processing, and the choice of statistical and data analysis technique used in this study are examined. The types of research designs adopted in this study include a longitudinal survey, a questionnaire survey and in-depth interviews. Major steps of the research designs are shown in Fig. 7.2. These methods are complementary to each other in order to gather information of and provide insights into the current issues and solutions.

References

Adams, G. B., & White, J. D. (1994). Dissertation research in public administration and cognate fields: An assessment of methods and quality. *Public Administration Review, 54*(6), 565–576.

Afonso, A., Schuknecht, L., & Tanzi, V. (2005). Public sector efficiency: An international comparison. *Public Choice, 123*(3–4), 321–347.

Alston, M., & Bowles, W. (1998). *Research for social workers: An introduction to methods*. St Leonards: Taylor & Francis.

Arvis, J.-F., Mustra, M. A., Panzer, J., Ojala, L., & Naula, T. (2007). *Connecting to compete: Trade logistics in the global*. Washington, D.C.: World Bank.

Borge, L.-E., Falch, T., & Tovmo, P. (2008). Public sector efficiency: the roles of political and budgetary institutions, fiscal capacity, and democratic participation. *Public Choice, 136*(3–4), 475–495.

Bowen, G. (2008). Naturalistic inquiry and the saturation concept: a research note. *Qualitative Research, 8*(1), 137.

Bryman, A., & Cramer, D. (2009). *Quantitative data analysis with SPSS 14, 15 and 16: A guide for social scientists*. New York: Routledge.

Burnard, P. (1991). A method of analysing interview transcripts in qualitative research. *Nurse Education Today, 11*(6), 461–466.

Carey, J. W. (1993). Linking qualitative and quantitative methods: Integrating cultural factors into public health. *Qualitative Health Research, 3*(3), 298.

Carley, K. (1993). Coding choices for textual analysis: A comparison of content analysis and map analysis. *Sociological Methodology, 23*, 75–126.

Crabtree, B. F., & Miller, W. L. (Eds.). (1999). *Doing qualitative research* (2nd ed.). Thousand Oaks, CA: Sage Publications Inc.

Creswell, J. W. (1998). *Qualitative inquiry and research design: choosing among five traditions*. Thousand Oaks, CA: Sage Publications.

Creswell, J. W., & Miller, D. L. (2000). Determining validity in qualitative inquiry. *Theory into Practice, 39*(3), 124–130.

Dollar, D., Micco, A., & Clark, X. (2002). *Maritime transport costs and port efficiency* (Policy Research Working Paper No. 2781). Washington, D.C.: World Bank.

Fellows, R. F., & Liu, A. M. (2003). *Research methods for construction* (2nd ed.). Malden, MA: Wiley-Blackwell.

Flyvbjerg, B. (2007). Megaproject policy and planning: problems causes, cures. *Environment and Planning B: Planning and Design, 34*, 578–597.

Geer, J. G. (1988). What do open-ended questions measure? *Public Opinion Quarterly, 52*(3), 365.

Glaser, B. G., & Strauss, A. L. (1967). *The discovery of grounded theory: Strategies for qualitative research*. London: Weidenfeld and Nicolson.

Gubrium, J. F., & Holstein, J. A. (2002). From the individual interview to the interview society. In J. F. Gubrium & J. A. Holstein (Eds.), *Handbook of interview research: Context and method* (pp. 3–32). Thousand Oaks, CA: Sage Publications Inc.

Guest, G., Bunce, A., & Johnson, L. (2006). How many interviews are enough? An experiment with data saturation and variability. *Field Methods, 18*(1), 59–82.

Hair, J. F., Black, W. C., & Babin, B. J. (2010). *Multivariate data analysis: A global perspective* (7th ed.). London: Pearson Education.

Hair, J. F., Ringle, C. M., & Sarstedt, M. (2011). PLS-SEM: Indeed a silver bullet. *The Journal of Marketing Theory and Practice, 19*(2), 139–152.

Hoang, V. (January 26, 2013). Nearly 70 % of firms in construction sector report making profit (in Vietnamese). *Dan Tri*. Retrieved from http://dantri.com.vn/kinh-doanh/gan-70-doanh-nghiep-nganh-xay-dung-bao-lai-689618.htm.

Hossain, L. (2000). Infrastructure development planning in SEA: An analysis of the national telecommunications planning practice in Thailand. *Technovation, 20*(5), 275–285.

Imriyas, K. (2007). A fuzzy knowledge-based system for premium-rating of workers' compensation insurance for construction projects (Doctoral dissertation, National University of Singapore, Singapore). Retrieved from http://scholarbank.nus.edu.sg/handle/10635/17612.

Jackson, S. L. (2006). *Research methods and statistics: A critical thinking approach* (2nd ed.). Belmont, CA: Thomson Wadsworth.

Jackson, K. M., & Trochim, W. M. K. (2002). Concept mapping as an alternative approach for the analysis of open-ended survey responses. *Organizational Research Methods, 5*(4), 307.

Johnson, R. B., & Onwuegbuzie, A. J. (2004). Mixed methods research: A research paradigm whose time has come. *Educational Researcher, 33*(7), 14.

Krippendorff, K. (1980). *Content analysis: An introduction to its methodology*. Beverly Hills: Sage.

Lan, Z., & Anders, K. K. (2000). A paradigmatic view of contemporary public administration research. *Administration and Society, 32*(2), 138.

Long, T., & Johnson, M. (2000). Rigour, reliability and validity in qualitative research. *Clinical Effectiveness in Nursing, 4*(1), 30–37.

Mandl, U., Dierx, A., & Ilzkovitz, F. (2008). *The effectiveness and efficiency of public spending*. Directorate General Economic and Monetary Affairs, European Commission.

Marshall, T. (2011). Reforming the process for infrastructure planning in the UK/England 1990–2010. *Town Planning Review, 82*(4), 441–467.

Mason, M. (2010). Sample size and saturation in Ph.D. studies using qualitative interviews. *Forum: Qualitative Social Research, 11*(3).

McCracken, G. D. (1988). *The long interview*. Thousand Oaks, CA: SAGE publications Inc.

McNabb, D. E. (2002). *Research methods in public administration and nonprofit management: Quantitative and qualitative approaches*. Armonk, NY: M.E. Sharpe.

Miller, G. J., & Whicker, M. L. (1999). *Handbook of research methods in public administration*. New York: Marcel Deckker.

Ministry of Construction. (2010). About us (in Vietnamese). Retrieved December 8, 2010, from http://www.xaydung.gov.vn/web/guest/gioi-thieu;jsessionid=02CA1415CC57A53E18CDC922AD80BBC5.

Ministry of Industry and Trade. (2010). Organizational structure (in Vietnamese). Retrieved December 28, 2010, from http://www.moit.gov.vn/vn/Pages/Tochuc.aspx.

Ministry of Industry and Trade. (2011). List of the most prominent traders selected and approved by the Ministry of Industry and Trade. Retrieved September 22, 2011, from http://dnxnk.moit.gov.vn/Categories.asp?Langs=1.

Ministry of Information and Communications. (2010). Organizational structure (in Vietnamese). Retrieved December 29, 2010, from http://mic.gov.vn/gioithieu/Trang/C%C6%A1c%E1%BA%A5ut%E1%BB%95ch%E1%BB%A9c.aspx.

Ministry of Information and Communications. (2011). List of leading firms in information and communications (in Vietnamese). Retrieved September 18, 2011, from http://mic.gov.vn/gioithieuSPDV/bc/Trang/dfdfdf.aspx.

Ministry of Natural Resources and Environment. (2010). Organizational structure (in Vietnamese). Retrieved December 30, 2010, from http://www.monre.gov.vn/v35/default.aspx?tabid=674&cateid=18.

Ministry of Planning and Investment. (2010). About us. Retrieved April 8, 2011, from http://www.mpi.gov.vn/portal/page/portal/bkhdt/gtcbkhdt.

Ministry of Transport. (2010). Organizational structure. Retrieved December 20, 2010, from http://www.mt.gov.vn/Default.aspx?tabid=47&catid=449.

Ministry of Transport. (2011). Data on firms (in Vietnamese). Retrieved September 26, 2011, from http://cucqlxd-bogtvt.vn/index.php?option=com_ntxl&view=list&Itemid=142.

Monecke, A., & Leisch, F. (2012). semPLS: Structural equation modelling using partial least squares. *Journal of Statistical Software, 48*(3), 1–32.

Morse, J. M. (1994). Designing funded qualitative research. In N. K. Denzin & Y. S. Lincoln (Eds.), *Handbook of qualitative research* (pp. 220–335). Thousand Oaks: Sage Publications.

Morse, J. (1995). The significance of saturation. *Qualitative Health Research, 5*(2), 147.

Morse, J. (2000). Determining sample size. *Qualitative Health Research, 10*(1), 3.

Mustajab, M. (2009). Infrastructure investment in Indonesia: Process and impact (Dissertation, University of Groningen, Netherlands). Retrieved from http://irs.ub.rug.nl/ppn/318943859.

Oppenheim, A. N. (1992). *Questionnaire design, interviewing and attitude measurement*. New York: Continuum.

Pang, G., & Herrera, S. (2005). Efficiency of public spending in developing countries: An efficiency frontier approach. World Bank Policy Research Working Paper (3645).

Reid, A. J. (1996). What we want: qualitative research—Promising frontier for family medicine. *Canadian Family Physician, 42*, 387–389.

Ritchie, J., Lewis, J., & Elam, G. (2003). Designing and selecting samples. In J. Ritchie & J. Lewis (Eds.), *Qualitative research practice: A guide for social science students and researchers* (pp. 77–108). Uk: Sage Publications Ltd.

Ryan, G. W., & Bernard, H. R. (2000). Data management and analysis methods. In N. K. Denzin & Y. S. Lincoln (Eds.), *Handbook of qualitative research* (2nd ed., pp. 769–802). London: Sage.

Sale, J. E. M., Lohfeld, L. H., & Brazil, K. (2002). Revisiting the quantitative-qualitative debate: Implications for mixed-methods research. *Quality and Quantity, 36*(1), 43–53.

Schramm, W. (1971) *Notes on case studies of instructional media project. Working paper for the academy for educational development*. Washington, D.C.

Schwab, K. (Ed.). (2008). *The global competitiveness report 2008–2009*. Geneva: World Economic Forum.

Schwab, K. (Ed.). (2009). *The global competitiveness report 2009–2010*. Geneva: World Economic Forum.

Seidel, J., & Kelle, U. (1995). Different functions of coding in the analysis of textual data. In U. Kelle, G. Prein, & K. Bird (Eds.), *Computer-aided qualitative data analysis: Theory, methods and practice* (pp. 52–61). London: Sage.

State Bank of Vietnam. (2011). List of financial institutions in Vietnam. Retrieved October 10, 2011, from http://www.sbv.gov.vn/portal/faces/vi/vilg/vilgpages_hethongtctd/nhtmnhanuoc?_adf.ctrl-state=ti5govu7y_825&_afrLoop=1258163460839100.

Stone, P. J. (1997). Thematic text analysis: New agendas for analyzing text content. In C. Robert (Ed.), *Text analysis for the social sciences: Methods for drawing statistical inferences from texts and transcripts* (pp. 35–54). Mahwah, N.J.: Lawrence Erlbaum.

Tan, W. C. K. (2007). *Practical research methods* (3rd ed.). Singapore: Prentice Hall.

Weber, R. P. (1990). *Basic content analysis*. Newbury Park, CA: Sage.

Webster, D., & Theeratham, P. (2004). Policy coordination, planning and infrastructure provision: A case study of Thailand: A background paper commissioned for the ADB-JBIC-World Bank East Asia and Pacific infrastructure flagship study.

Wilson, J. S., Mann, C. L., & Otsuki, T. (2003). *Trade facilitation and economic development: Measuring the impact* (Policy Research Working Paper No. 2988). Washington, D.C.: World Bank.

Wilson, J. S., Soloaga, I., & Mejia, A. (2006). *Moving forward faster: trade facilitation reform and Mexican competitiveness* (Policy Research Working Paper No. 3953). Washington, D.C.: World Bank.

Yin, R. K. (2009). *Case study research: Design and methods* (4th ed.). Los Angeles, CA: Sage Publications Inc.

Chapter 8
Data Analysis and Discussions

Abstract This chapter analyses data collected and discusses the results. The chapter starts with a study of the efficiency of public investments in infrastructure in Vietnam. Next, a model of the critical factors affecting the current use of public investments in infrastructure to support trade and economic growth in Vietnam is examined and validated. From the data analysis, four groups of factors emerge as key challenges in moving towards the efficient use of public investments in infrastructure, including inadequate forecasting capacity, lack of transparency and accountability, lack of an effective land acquisition framework, insufficient building capacity of local firms, and insufficient institutional capacity. Results from both quantitative and qualitative data analyses are discussed and explained together in the last section to enrich the research findings.

Keywords Critical factors · Institutional capacity of the government · Forecasting capacity · Transparency and accountability · Land acquisition · Building capacity of local firms

8.1 Introduction

This chapter analyses data collected and discusses the results. The chapter starts with a study of the efficiency of public investments in infrastructure. Next, the survey data are analysed, and hypotheses are tested using several statistical techniques. Results obtained from the survey are discussed and explained together with the interviews in the last section. The research framework is expected to emerge from the discussion.

G. Dang and L. Sui Pheng, *Infrastructure Investments in Developing Economies*,
DOI 10.1007/978-981-287-248-7_8

8.2 Longitudinal Analysis

The purpose of the analysis is to provide an overview of the level of efficiency of public investments in infrastructure to support trade and economic growth in Vietnam. Notably, the analysis does not attempt to formulate regression models to establish relationship between public investments in infrastructure and the ability of infrastructure to support trade and economic growth. Rather, it seeks to provide evidence that reflects the efficiency of public investments in infrastructure to support trade and economic growth over time. As explained in Sect. 7.4, the efficiency of public investments in infrastructure to support trade and economic growth is measured by weighting the ability of infrastructure to support trade and economic growth by the amount of public investments in infrastructure used to achieve the ability (Afonso et al. 2005):

$$E = \frac{A}{F}$$

where

E Efficiency of public investments in infrastructure to support trade and economic growth;

A Ability of infrastructure to support trade and economic growth; and

F Public investments in infrastructure (as % of GDP)

The ability of infrastructure to support trade and economic growth (A) is measured by the competitiveness index of Vietnam in terms of infrastructure development as reported in the Global Competitiveness Reports published annually by the World Economic Forum (WEF). F is the public expenditure used to achieve the competeveness level of Vietnam in terms of infrastructure development. Government funding invested in infrastructure construction (F) is measured by state investments (as % of GDP) in electricity supply, transport, and communications as reported in the statistical yearbooks of Vietnam. Collected data are given in Table 8.1.

Due to the time lag characteristic of construction outputs (Ofori 1990), a lagged effect from investment on competitiveness is assumed in order to compute the efficiency indicator. As mentioned in Sect. 6.6.2.1, the five-year Socio-economic Development Plan of Vietnam is accompanied by a public investment program (PIP), which serves as the basis for the formulation of infrastructure investments (Pincus and Nguyen 2004). Therefore, a time lag of five years is applicable in this study. For example, public investments in infrastructure in 1995 are assumed to affect the competitiveness of infrastructure in 1999. For example,

Table 8.1 Efficiency of public investments in infrastructure construction to support trade and economic growth

Year	Ability of infrastructure to support trade and economic growth (A)[a]	Government investments in infrastructure construction (F as % of GDP)[b] (%)	Efficiency indicator $E = A/F^c$	$\Delta E = \Delta (A/F)$	ΔF^c (%)	Marginal productivity of government funding	
						$\partial E/\partial F = \Delta E/\Delta F$	$(\partial E/\partial F)/100$
1995		5.65					
1996		7.11					
1997	2.08	7.72					
1998	3.22	7.85					
1999	2.63	8.13	46.54				
2000	2.1	7.81	29.52	−17.02	1.46	−1,163.58	−11.64
2001	2.2	7.74	28.51	−1.01	0.60	−167.53	−1.68
2002	2.5	8.48	31.85	3.34	0.13	2,515.50	25.16
2003	2.7	7.62	33.20	1.35	0.28	478.31	4.78
2004	2.6	7.84	33.29	0.09	−0.32	−27.94	−0.28
2005	2.6	7.93	33.61	0.32	−0.07	−430.41	−4.30
2006	2.6	7.41	30.66	−2.96	0.75	−396.32	−3.96
2007	2.8	7.05	36.76	6.10	−0.86	−706.52	−7.07
2008	2.86	6.21	36.48	−0.28	0.22	−124.13	−1.24
2009	3	7.68	37.85	1.37	0.09	1,583.94	15.84
2010	3.56	7.07	48.02	10.17	−0.51	−1,985.28	−19.85
			50.89	2.87	−0.36	−799.00	−7.99

[a] Data collected from Global Competitiveness Reports issued by WEF from 1997 to 2010
[b] Data collected from Statistical Yearbooks issued by GSO from 1999 to 2010
[c] A 5-year lagged effect from government funding is applied

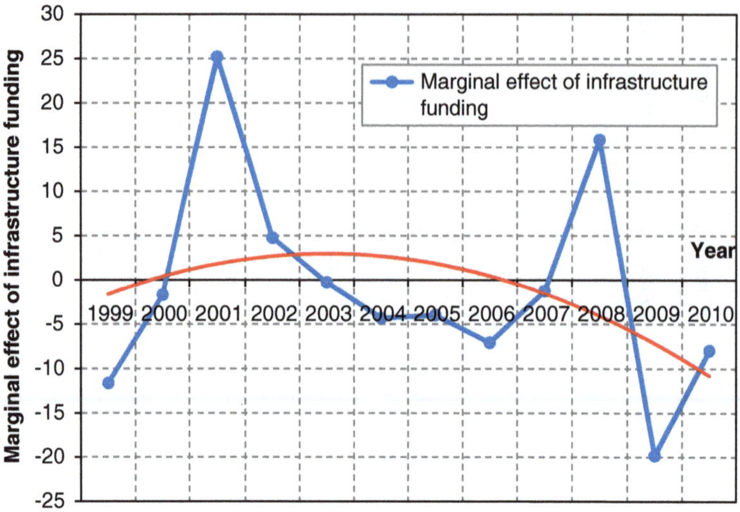

Fig. 8.1 Marginal effect of public investments in infrastructure for supporting trade and economic growth in Vietnam

$$E_{1999} = A_{1999}/F_{1995} = 2.63/5.65\,\% = 46.54$$
$$\Delta F_{2000} = F_{1996} - F_{1995} = 7.11\,\% - 5.65\,\% = 1.46\,\%$$
$$\Delta E_{2000} = F_{2000} - F_{1999} = 9.52 - 46.54 = -17.2.$$

The results shown in Table 8.1 indicated that efficiency of public investments in infrastructure increases across the years, i.e. $\Delta E = \Delta(A/F) > 0$ (except in 2000, 2001, 2006, and 2008). To extract more information from the efficiency indicator, marginal effect of the public investments in infrastructure is computed. The marginal effect ($\partial E/\partial F$) is measured by the impact on efficiency of one-unit change in public investments in infrastructure. The marginal effect was divided by 100 for easy reference (see the last column in Table 8.1). As shown in Fig. 8.1, the marginal effect peaked in 2001 and then went down from 2002 to 2007. The marginal effect peaked again in 2008 and then went down in 2009. The overall trend of the marginal effect (see red line in Fig. 8.1) presents a declining trend. In other words, the efficiency of public investments in infrastructure to support trade and economic growth is falling. Since the marginal effect indicator has taken a weighted average of capital investment into consideration (i.e. one-unit change in public investments in infrastructure), the declining trend implies that there are issues other than capital provision reducing the marginal effect. The efficiency problem implies that further study on what causal factors are and how to overcome them are necessary in Vietnam.

8.3 Factor Analysis

8.3.1 Sample Profile

The survey was conducted in Vietnam between March and June 2012. Questionnaires were directly sent to professionals in Hanoi and sent via email to professionals in Ho-Chi-Minh City (see Sect. 7.5.1 for sampling method). In total, 319 questionnaires were sent out. There were 126 respondents returning their responses. The response rate is thus about 39 %. The sample size is adequate for meaningful statistical analysis (Hair et al. 2010). Table 8.2 provides the characteristics of the respondents. Out of 126 respondents, 39 respondents, accounting for 31 %, were from government agencies. Most of them are advisors, senior advisors, and researchers. Two department heads and three department deputies participated in the survey. The remaining respondents consist of developers, consultants/contractors, traders, and financial institutions, accounting for 69 % of total respondents. Their professional and management levels are provided in Table 8.3. Out of 126 respondents, 16 respondents answered the last open-ended question about further issues other than those mentioned in the questionnaires by the researcher. The issues mentioned in these answers have been more or less captured in the questionnaires. As a result, these answers did not provide useful information for further understanding of issues under consideration. These answers were thus not used for further analysis.

8.3.2 Ranking of Problems' Occurrence and Criticality

From the literature review, 35 problems that would affect the efficient use of public investments in infrastructure were identified. They were approximately divided into the categories shown in Table 8.4.

Table 8.2 Questionnaire return rate

Group	Questionnaires distributed	Responses received	Response rate (%)	Proportion (%)
Government agencies	**108**	**39**	**36**	**31**
Non-government agencies	**211**	**87**	**41**	**69**
Developers	19	19		
Consultants/ contractors	112	31		
Traders	37	16		
Financial institutions	43	21		
Total	**319**	**126**	**39**	**100**

Bolds are the total number of respondents of different groups

Table 8.3 Type of management levels

Group	Upper management	Middle management	Professionals	Total
Government agencies	5	15	19	39
Percentage	**13 %**	**38 %**	**49 %**	**100 %**
Non-government agencies	21	38	28	87
Developers	9	5	5	19
Consultants/contractors	8	17	6	31
Traders	2	9	5	16
Financial institutions	2	7	12	21
Percentage	**24 %**	**44 %**	**32 %**	**100 %**

Bolds are the percentage of respondents that belong to a certain type of management levels within different groups

The full rankings of problems in terms of their occurrence and criticality by non-government agencies and by government agencies are given in Tables 8.5 and 8.6. The rankings are based on sample median occurrence rating and median critical rating, using the following formula (Tan 2007):

$$m = \frac{1}{n}\sum x_i$$

where m is the sample median value, n is the sample size (n = 39 for the government group and n = 87 for the non-government group), x_i is the score rated by respondent i. For example, the median occurrence rating of X_{15} is computed as follows:

$$\begin{aligned} m(\text{of } X_{15} \text{ by the non-government group}) &= \frac{1}{87}(x_1 + x_2 + \cdots + x_{87}) \\ &= \frac{1}{87}(4 + 3 + 4 + \cdots + 3 + 3 + 2) = 3.74. \end{aligned}$$

A similar method is applied for computing sample median critical rating. Results are provided in Tables 8.5 and 8.6.

From the ratings, many problems were ranked highly by both non-government and government actors. Examples are:

- Complicated and bureaucratic administrative system in infrastructure development planning (X_{15});
- Inadequate capacity of regulatory bodies to enforce regulations to reduce corruption in funded infrastructure construction projects (X_{23});
- Bureaucracy in settling land disputes and claims (X_{28});

Table 8.4 List of variables affecting the efficient use of public investments in infrastructure

Categories	Variables	
Capacity for estimating and monitoring of rates of return of infrastructure projects	Limitations of forecasting methods and appraisal techniques	X_1
	Inadequate data	X_2
	Inherent problems in predicting the future and monetizing external and indirect effects	X_3
	Lack of experienced forecasters	X_4
	Inadequacy in ex-post analysis on whether projects meet objectives	X_5
Politicized decision-making	Political leaders and government bureaucrats make decisions in their personal, sectoral, or regional interests (e.g. securing political positions or competing for scarce funds) as opposed to national goals	X_6
	Interest group pressure	X_7
Transparency and accountability	Lack of consultation with the interested or affected individuals or groups about forecasting methods, criteria of project selection and the determination of planning objectives	X_8
	Lack of independent peer reviews and quality checks on forecasts and planning outcomes by independent-review bodies and scientific and professional community	X_9
	Lack of penal systems to enforce penalties on those that deliberately and consistently produce deceptive forecasts	X_{10}
Institutional weaknesses in planning	Poor communication between planning agency and the day-to-day decision-making machinery of government	X_{11}
	Lack of inter-ministerial and intersectoral coordination	X_{12}
	Lack of interaction between political leaders and planners with non-governmental actors	X_{13}
	Incompetent and unqualified civil servants	X_{14}
	Complicated and bureaucratic administrative system	X_{15}
Political commitment	Lack of political will on the part of top leaders and high-level decision-makers	X_{16}
	Changes in the political affiliations of the government responsible for the projects	X_{17}
	Political inconsistencies at national level and between different tiers of governments (central, regional, and local)	X_{18}
	Lack of a high-powered government institution that provides an effective mechanism for implementing national infrastructure plans	X_{19}

(continued)

Table 8.4 (continued)

Categories	Variables	
Corruption in infrastructure construction	Lack of transparency and competitiveness in bid processes	X_{20}
	Discretionary power of individual bureaucrats involved in award of contracts	X_{21}
	Inadequate financial and physical auditing	X_{22}
	Inadequate capacity of regulatory bodies to enforce regulations	X_{23}
Building capacity of local firms	Poor level of efficiency and quality of work	X_{24}
	Poor level of professionalism and entrepreneurship	X_{25}
	Resources shortages, especially in construction technology, management, and finance	X_{26}
Land acquisition	Lack of a negotiating mechanism to make land acquisition compensation more market-oriented	X_{27}
	Bureaucracy in settling land disputes and claims	X_{28}
	Lack of a land acquisition compensation-monitoring system	X_{29}
	Lack of clarity about valuation methods for compensation	X_{30}
	Lack of law enforcement to regulate the price of land acquired for infrastructure projects	X_{31}
Institutional and legal weaknesses in infrastructure construction	Obsolescence of building regulations	X_{32}
	Changing and inconsistent law and regulations	X_{33}
	Ineffectiveness of implementation of existing statutes and codes	X_{34}
	Bureaucracy in formal procedures relating to infrastructure project planning, construction permission, and administration	X_{35}

- Lack of a negotiating mechanism to make land acquisition compensation more market-oriented (X_{27});
- Poor level of efficiency and quality of work of local construction firms (X_{24}); and
- Lack of clarity about valuation methods for compensation for land acquired for infrastructure projects (X_{30}).

However, in order to ascertain the degree of agreement between the perceptions of the non-government-agency group and government-agency group towards these problems, the Spearman rank correlation coefficient r is computed using the bivariate correlation procedures from SPSS. The Spearman rank correlation

Table 8.5 Median occurrence rating of problems

Ranking	Non-government agencies		Government agencies	
	Problems	Median occurrence rating	Problems	Median occurrence rating
1	X_{15}	3.74	X_{4}	3.23
2	X_{23}	3.60	X_{10}	3.21
3	X_{35}	3.60	X_{28}	3.21
4	X_{28}	3.57	X_{26}	3.15
5	X_{20}	3.44	X_{30}	3.15
6	X_{30}	3.25	X_{24}	3.13
7	X_{10}	3.24	X_{27}	3.10
8	X_{24}	3.22	X_{29}	3.10
9	X_{29}	3.22	X_{3}	3.08
10	X_{5}	3.18	X_{15}	3.03
11	X_{25}	3.18	X_{5}	2.97
12	X_{27}	3.16	X_{23}	2.97
13	X_{14}	3.14	X_{25}	2.97
14	X_{26}	3.07	X_{2}	2.92
15	X_{7}	3.06	X_{9}	2.90
16	X_{31}	3.05	X_{31}	2.90
17	X_{4}	3.03	X_{35}	2.77
18	X_{9}	3.02	X_{7}	2.74
19	X_{19}	2.99	X_{14}	2.74
20	X_{6}	2.93	X_{34}	2.74
21	X_{33}	2.91	X_{20}	2.72
22	X_{3}	2.90	X_{22}	2.69
23	X_{22}	2.89	X_{1}	2.62
24	X_{21}	2.87	X_{8}	2.56
25	X_{2}	2.82	X_{12}	2.54
26	X_{1}	2.76	X_{19}	2.44
27	X_{12}	2.74	X_{6}	2.38
28	X_{13}	2.74	X_{11}	2.38
29	X_{34}	2.72	X_{13}	2.36
30	X_{8}	2.59	X_{33}	2.33
31	X_{32}	2.55	X_{21}	2.21
32	X_{18}	2.37	X_{18}	2.15
33	X_{16}	2.33	X_{32}	2.15
34	X_{11}	2.30	X_{16}	2.10
35	X_{17}	2.15	X_{17}	2.08

Table 8.6 Median criticality rating of problems

Ranking	Non-government agencies		Government agencies	
	Problems	Median criticality rating	Problems	Median criticality rating
1	X_{23}	3.97	X_{26}	3.67
2	X_{35}	3.92	X_{24}	3.62
3	X_{15}	3.90	X_{23}	3.59
4	X_{28}	3.83	X_{25}	3.51
5	X_{20}	3.77	X_{29}	3.44
6	X_{24}	3.68	X_{20}	3.41
7	X_{14}	3.62	X_{30}	3.38
8	X_{30}	3.59	X_{31}	3.38
9	X_{27}	3.55	X_{27}	3.36
10	X_{25}	3.49	X_{15}	3.36
11	X_{31}	3.48	X_{35}	3.36
12	X_{29}	3.46	X_{28}	3.28
13	X_{10}	3.45	X_7	3.26
14	X_7	3.37	X_{14}	3.23
15	X_{19}	3.37	X_{22}	3.21
16	X_{22}	3.36	X_4	3.15
17	X_{26}	3.34	X_{10}	3.15
18	X_4	3.34	X_3	3.05
19	X_6	3.31	X_{34}	3.03
20	X_{33}	3.31	X_{33}	3.03
21	X_{21}	3.31	X_2	3.00
22	X_5	3.23	X_9	3.00
23	X_2	3.21	X_6	3.00
24	X_9	3.20	X_{16}	3.00
25	X_{12}	3.13	X_5	2.97
26	X_{34}	3.09	X_8	2.97
27	X_3	3.01	X_{19}	2.95
28	X_1	2.98	X_{21}	2.87
29	X_{13}	2.98	X_{32}	2.85
30	X_{16}	2.97	X_{12}	2.79
31	X_8	2.82	X_1	2.77
32	X_{32}	2.80	X_{18}	2.77
33	X_{18}	2.79	X_{13}	2.72
34	X_{11}	2.76	X_{11}	2.69
35	X_{17}	2.47	X_{17}	2.46

coefficients for ranking of the occurrence of problems and of the criticality of problems between the two groups are 0.748 and 0.844 (with $p = 0.00$), respectively. With $p < 0.01$, there was a strong evidence to believe that the correlations are significant at the 1 % significance level. The statistically significant correlation thus implies that there is a high degree of agreement between the two groups on the level of occurrence and criticality of problems. Thus, separate factor analyses for each group are not necessary to be performed (Hair et al. 2010).

Among 35 problems, only problems with their degree of criticality perceived from moderate to very significant are selected for factor analysis in the following section. These problems thus should have their median of criticality from 3 to 5 on a scale of 1–5. As a result, only the top ranking problems with the median critical ratings from 3 from the group of respondents combining both non-government and government agencies are selected for factor analysis. In other words, the median difference between these problems and "3" should be positive (Table 8.7). One-tailed hypothesis test on the median of criticality of these problems was used with test value of 3. The null hypothesis is H_0: $M = 3$ and the alternative hypothesis H_1: $M < 3$. The test statistic is (Tan 2007):

$$t = \frac{m - 3}{s/\sqrt{n}}$$

where

m sample median
s sample variance
n sample size

Only problems with positive values of median difference (between the observed median and the test value of 3) and t-values higher than the critical value at the 5 % significance level are selected as critical factors for factor analysis. The test results on the median of critical ratings for each problem are computed using the one-sample t-test procedures in SPSS. The test results given in Table 8.7 show that the first 27 top ranked problems have positive values of median difference and the t-values of these problems are higher than the critical value at the 5 % significance level (−1.65). That means there is no evidence to reject H_0 and accept H_1. Therefore, it can be concluded that at the 0.05 level of significance, there is no evidence to support the proposition that the median criticality of these top 27 ranked problems is less than 3 (Table 8.7).

The remaining 8 problems were identified as unimportant by non-government and government agencies in infrastructure development in Vietnam, including:

- Limitations of forecasting methods and appraisal techniques (X_1);
- Lack of consultation with the interested or affected individuals or groups about forecasting methods, criteria of project selection and the determination of planning objectives (X_8);

Table 8.7 Hypothesis test on the median of criticality ratings

One-sample statistics					Test value = 3			95 % confidence interval of the difference	
Rank		Median	Std. Dev.	Std. error mean	t	df	Median difference	Lower	Upper
1	X_{23}	3.85	1.066	0.095	8.939	125	0.849	0.66	1.04
2	X_{35}	3.75	0.987	0.088	8.481	125	0.746	0.57	0.92
3	X_{15}	3.73	0.933	0.083	8.784	125	0.730	0.57	0.89
4	X_{20}	3.66	0.948	0.084	7.800	125	0.659	0.49	0.83
5	X_{24}	3.66	0.878	0.078	8.423	125	0.659	0.50	0.81
6	X_{28}	3.66	1.021	0.091	7.242	125	0.659	0.48	0.84
7	X_{30}	3.52	1.033	0.092	5.691	125	0.524	0.34	0.71
8	X_{14}	3.50	1.101	0.098	5.098	125	0.500	0.31	0.69
9	X_{25}	3.50	0.927	0.083	6.052	125	0.500	0.34	0.66
10	X_{27}	3.49	0.978	0.087	5.649	125	0.492	0.32	0.66
11	X_{29}	3.45	0.985	0.088	5.157	125	0.452	0.28	0.63
12	X_{31}	3.45	1.001	0.089	5.074	125	0.452	0.28	0.63
13	X_{26}	3.44	0.872	0.078	5.719	125	0.444	0.29	0.60
14	X_{10}	3.36	0.975	0.087	4.110	125	0.357	0.19	0.53
15	X_7	3.33	1.066	0.095	3.511	125	0.333	0.15	0.52
16	X_{22}	3.31	1.039	0.093	3.344	125	0.310	0.13	0.49
17	X_4	3.29	1.011	0.090	3.173	125	0.286	0.11	0.46
18	X_{19}	3.24	1.099	0.098	2.433	125	0.238	0.04	0.43
19	X_{33}	3.22	1.019	0.091	2.448	125	0.222	0.04	0.40
20	X_6	3.21	1.211	0.108	1.987	125	0.214	0.00	0.43
21	X_{21}	3.17	0.989	0.088	1.983	125	0.175	0.00	0.35
22	X_5	3.15	1.089	0.097	1.555	125	0.151	−0.04	0.34
23	X_2	3.14	1.056	0.094	1.518	125	0.143	−0.04	0.33
24	X_9	3.13	0.950	0.085	1.595	125	0.135	−0.03	0.30
25	X_{34}	3.07	0.989	0.088	0.810	125	0.071	−0.10	0.25
26	X_3	3.02	0.907	0.081	0.295	125	0.024	−0.14	0.18
27	X_{12}	3.02	1.000	0.089	0.267	125	0.024	−0.15	0.20
28	X_{16}	2.98	1.039	0.093	−0.257	125	−0.024	−0.21	0.16
29	X_1	2.91	0.938	0.084	−1.044	125	−0.087	−0.25	0.08
30	X_{13}	2.90	0.875	0.078	−1.324	125	−0.103	−0.26	0.05
31	X_8	2.87	0.880	0.078	−1.722	125	−0.135	−0.29	0.02
32	X_{32}	2.82	1.061	0.095	−1.931	125	−0.183	−0.37	0.00
33	X_{18}	2.79	1.100	0.098	−2.187	125	−0.214	−0.41	−0.02
34	X_{11}	2.74	0.981	0.087	−2.996	125	−0.262	−0.43	−0.09
35	X_{17}	2.47	1.001	0.089	−5.960	125	−0.532	−0.71	−0.36

- Poor communication between planning agency and the day-to-day decision-making machinery of government (X_{11});
- Lack of interaction between political leaders and planners with non-governmental actors (X_{13});
- Lack of political will on the part of top leaders and high-level decision-makers (X_{16});
- Changes in the political affiliations of the government responsible for the projects (X_{17});
- Political inconsistencies at national level and between different tiers of governments (central, regional, and local) (X_{18}); and
- Obsolescence of building regulations (X_{32}).

Under "Capacity for estimating and monitoring rates of return" category, limitations of forecasting methods and appraisal techniques (X_1) were considered as not having critical impact on the efficiency of government-funded infrastructure construction in Vietnam. This could be explained that forecasting methods and techniques are often expected to be refined over time. Better forecasting methods and techniques in national infrastructure development plans would have been developed (Flyvbjerg 2007). Forecasting techniques are thus not a major concern among respondents.

Under "Institutional weaknesses in planning" category, poor communication between planning agency and the day-to-day decision-making machinery of government (X_{11}) and lack of interaction between political leaders and planners with non-governmental actors (X_{13}) have somewhat critical influence on the efficiency of government-funded infrastructure construction in Vietnam. As described in Chap. 6 , the planning process involved a wide range of actors. This implies that there are links between planning agency and other part of government machinery, and non-governmental actors as well. As a result, X_{11} and X_{13} were not important factors for further analysis.

Problems X_{16}, X_{17}, and X_{18} under the "Political commitment" category are eliminated from further factor analysis. This can be explained that having only one party, the Communist Party of Vietnam, since independence has ensured the consistency of the political system in Vietnam. With all senior and top government positions occupied by party members, decisions made are committed to the direction of the Party (Arkadie and Mallon 2003).

Finally, "Obsolescence of building regulations" (X_{32}) is not identified as one of the major issues affecting the efficiency of government-funded infrastructure construction in Vietnam. This suggests that a number of updates of appropriate standards, codes, and regulations in recent years have eased the concern about obsolescence. The application of relevant technical standards such as ISO, British, American, French, and Japanese standards in Vietnam have also helped to improve the industry practices (Chiang et al. 2004).

8.3.3 Factor Analysis of Problems' Occurrence

8.3.3.1 Assumptions in Factor Analysis

As mentioned earlier, the top 27 high-ranked problems in terms of their criticality were selected for factor analysis. The purpose of this analysis is to capture the multivariate relationship existing among problems in terms of their degree of occurrence. All measured variables are metric. The sample constitutes a homogeneous set of perceptions. The sample size of 126 provides an adequate basis for the computation of the correlations between variables (Hair et al. 2010).

Several tests are required to justify the application of factor analysis. The purpose of these tests, including visual inspection, the Bartlett test of sphericity, and the measure of sampling adequacy (MSA), is to ensure that the data matrix has sufficient correlations for factor analysis to proceed (Hair et al. 2010). The test results are given in Tables 8.8 and 8.9 using factor analysis procedures in SPSS.

Visual inspection of the correlations among the 27 variables reveals a substantial number of correlations greater than 0.30, and all correlations are significant at 5 % level (see Table 8.8). The partial correlations also show no value greater than 0.5, indicating that factor analysis is appropriate (see Table 8.8). The value of the Bartlett test of sphericity is 2,275.5, and the correlation matrix has significant correlations. The overall MSA value is 0.91, and the MSA value for each variable is more than 0.8, meeting the fundamental requirements for factor analysis (see Table 8.9). Overall, the results of these tests suggest that a structure exists to group variables, and thus, the next steps of factor analysis are applicable (Hair et al. 2010).

8.3.3.2 Choosing Factor Models and Number of Factors

With p variables and k common factors to be determined and a unique factor δ, the factor analytic model can be written as (Tan 2007):

$$\begin{aligned} X_1 &= l_{11}F_1 + l_{12}F_2 + \cdots + l_{1k}F_k + \delta_1 \\ X_2 &= l_{21}F_1 + l_{22}F_2 + \cdots + l_{2k}F_k + \delta_2 \\ &\ldots \\ X_p &= l_{p1}F_1 + l_{p2}F_2 + \cdots + l_{pk}F_k + \delta_p \end{aligned}$$

In this study, $p = 27$ and F_s are common factors to be determined. The coefficients (l_s) are called factor loadings, since they indicate the degree of association of each factor to each variable. The process of estimation factor loadings starts with determining the number of factor to be retained. Principle component analysis was used to extract factors. The results are shown in Table 8.10.

According to the latent root criterion, only the factors having eigenvalues greater than 1 are considered to be retained. As shown in Table 8.10, there are five factors

Table 8.8 Correlation matrix

Variables		X2	X3	X4	X5	X6	X7	X9	X10	X12	X14	X15	X19	X20	X21
Correlation	X2	1.00	0.45	0.49	0.49	0.39	0.47	0.46	0.40	0.55	0.36	0.30	0.37	0.33	0.27
	X3	0.45	1.00	0.46	0.47	0.37	0.39	0.41	0.34	0.35	0.26	0.22	0.24	0.23	0.22
	X4	0.49	0.46	1.00	0.48	0.39	0.45	0.53	0.39	0.47	0.42	0.35	0.29	0.33	0.22
	X5	0.49	0.47	0.48	1.00	0.50	0.49	0.58	0.43	0.56	0.50	0.47	0.57	0.39	0.27
	X6	0.39	0.37	0.39	0.50	1.00	0.71	0.53	0.43	0.55	0.63	0.47	0.41	0.53	0.37
	X7	0.47	0.39	0.45	0.49	0.71	1.00	0.54	0.41	0.52	0.55	0.50	0.39	0.53	0.34
	X9	0.46	0.41	0.53	0.58	0.53	0.54	1.00	0.54	0.58	0.53	0.38	0.52	0.39	0.33
	X10	0.40	0.34	0.39	0.43	0.43	0.41	0.54	1.00	0.44	0.49	0.42	0.34	0.38	0.32
	X12	0.55	0.35	0.47	0.56	0.55	0.52	0.58	0.44	1.00	0.61	0.54	0.49	0.54	0.44
	X14	0.36	0.26	0.42	0.50	0.63	0.55	0.53	0.49	0.61	1.00	0.59	0.45	0.59	0.44
	X15	0.30	0.22	0.35	0.47	0.47	0.50	0.38	0.42	0.54	0.59	1.00	0.41	0.57	0.39
	X19	0.37	0.24	0.29	0.57	0.41	0.39	0.52	0.34	0.49	0.45	0.41	1.00	0.47	0.41
	X20	0.33	0.23	0.33	0.39	0.53	0.53	0.39	0.38	0.54	0.59	0.57	0.47	1.00	0.72
	X21	0.27	0.22	0.22	0.27	0.37	0.34	0.33	0.32	0.44	0.44	0.39	0.41	0.72	1.00
	X22	0.46	0.44	0.44	0.47	0.53	0.53	0.62	0.54	0.58	0.62	0.46	0.49	0.61	0.48
	X23	0.36	0.44	0.39	0.56	0.48	0.45	0.45	0.52	0.53	0.61	0.61	0.44	0.67	0.54
	X24	0.27	0.24	0.26	0.37	0.41	0.45	0.38	0.42	0.28	0.49	0.33	0.31	0.41	0.33
	X25	0.33	0.44	0.37	0.42	0.48	0.51	0.48	0.35	0.37	0.56	0.38	0.38	0.46	0.38
	X26	0.36	0.39	0.48	0.41	0.43	0.45	0.42	0.40	0.30	0.50	0.32	0.33	0.41	0.37
	X27	0.47	0.41	0.49	0.44	0.42	0.42	0.40	0.35	0.47	0.34	0.35	0.46	0.52	0.44
	X28	0.33	0.33	0.35	0.46	0.44	0.46	0.31	0.33	0.48	0.49	0.52	0.36	0.61	0.35
	X29	0.31	0.25	0.38	0.36	0.37	0.44	0.36	0.33	0.43	0.40	0.33	0.36	0.43	0.25
	X30	0.42	0.31	0.34	0.48	0.44	0.55	0.46	0.36	0.47	0.42	0.42	0.37	0.58	0.37
	X31	0.42	0.36	0.28	0.55	0.39	0.38	0.52	0.39	0.52	0.47	0.36	0.49	0.43	0.20
	X33	0.37	0.19	0.21	0.38	0.47	0.44	0.45	0.38	0.44	0.61	0.49	0.40	0.55	0.42
	X34	0.54	0.49	0.57	0.61	0.50	0.52	0.62	0.49	0.61	0.45	0.49	0.50	0.53	0.42
	X35	0.36	0.32	0.39	0.46	0.45	0.52	0.44	0.37	0.47	0.46	0.66	0.38	0.66	0.49

(continued)

Table 8.8 (continued)

Variables		X22	X23	X24	X25	X26	X27	X28	X29	X30	X31	X33	X34	X35
Correlation	X2	0.46	0.36	0.27	0.33	0.36	0.47	0.33	0.31	0.42	0.42	0.37	0.54	0.36
	X3	0.44	0.44	0.24	0.44	0.39	0.41	0.33	0.25	0.31	0.36	0.19	0.49	0.32
	X4	0.44	0.39	0.26	0.37	0.48	0.49	0.35	0.38	0.34	0.28	0.21	0.57	0.39
	X5	0.47	0.56	0.37	0.42	0.41	0.44	0.46	0.36	0.48	0.55	0.38	0.61	0.46
	X6	0.53	0.48	0.41	0.48	0.43	0.42	0.44	0.37	0.44	0.39	0.47	0.50	0.45
	X7	0.53	0.45	0.45	0.51	0.45	0.42	0.46	0.44	0.55	0.38	0.44	0.52	0.52
	X9	0.62	0.45	0.38	0.48	0.42	0.40	0.31	0.36	0.46	0.52	0.45	0.62	0.44
	X10	0.54	0.52	0.42	0.35	0.40	0.35	0.33	0.33	0.36	0.39	0.38	0.49	0.37
	X12	0.58	0.53	0.28	0.37	0.30	0.47	0.48	0.43	0.47	0.52	0.44	0.61	0.47
	X14	0.62	0.61	0.49	0.56	0.50	0.34	0.49	0.40	0.42	0.47	0.61	0.45	0.46
	X15	0.46	0.61	0.33	0.38	0.32	0.35	0.52	0.33	0.42	0.36	0.49	0.49	0.66
	X19	0.49	0.44	0.31	0.38	0.33	0.46	0.36	0.36	0.37	0.49	0.40	0.50	0.38
	X20	0.61	0.67	0.41	0.46	0.41	0.52	0.61	0.43	0.58	0.43	0.55	0.53	0.66
	X21	0.48	0.54	0.33	0.38	0.37	0.44	0.35	0.25	0.37	0.20	0.42	0.42	0.49
	X22	1.00	0.58	0.38	0.57	0.50	0.41	0.36	0.32	0.42	0.50	0.48	0.58	0.46
	X23	0.58	1.00	0.42	0.44	0.50	0.47	0.55	0.38	0.44	0.46	0.45	0.51	0.68
	X24	0.38	0.42	1.00	0.67	0.52	0.37	0.35	0.28	0.45	0.37	0.46	0.30	0.33
	X25	0.57	0.44	0.67	1.00	0.61	0.40	0.38	0.35	0.53	0.45	0.50	0.53	0.44
	X26	0.50	0.50	0.52	0.61	1.00	0.36	0.33	0.26	0.39	0.34	0.39	0.48	0.38
	X27	0.41	0.47	0.37	0.40	0.36	1.00	0.59	0.45	0.58	0.54	0.28	0.60	0.45
	X28	0.36	0.55	0.35	0.38	0.33	0.59	1.00	0.50	0.64	0.55	0.39	0.49	0.61
	X29	0.32	0.38	0.28	0.35	0.26	0.45	0.50	1.00	0.68	0.61	0.29	0.48	0.43
	X30	0.42	0.44	0.45	0.53	0.39	0.58	0.64	0.68	1.00	0.63	0.42	0.58	0.52
	X31	0.50	0.46	0.37	0.45	0.34	0.54	0.55	0.61	0.63	1.00	0.37	0.50	0.42
	X33	0.48	0.45	0.46	0.50	0.39	0.28	0.39	0.29	0.42	0.37	1.00	0.44	0.48
	X34	0.58	0.51	0.30	0.53	0.48	0.60	0.49	0.48	0.58	0.50	0.44	1.00	0.59
	X35	0.46	0.68	0.33	0.44	0.38	0.45	0.61	0.43	0.52	0.42	0.48	0.59	1.00

(continued)

Table 8.8 (continued)

Variables		X_2	X_3	X_4	X_5	X_6	X_7	X_9	X_{10}	X_{12}	X_{14}	X_{15}	X_{19}	X_{20}	X_{21}
Sig. (1-tailed)	X_2		0.00	0.00	0.00	0.00	0.00	0.00	0.00	0.00	0.00	0.00	0.00	0.00	0.00
	X_3	0.00		0.00	0.00	0.00	0.00	0.00	0.00	0.00	0.00	0.01	0.00	0.00	0.01
	X_4	0.00	0.00		0.00	0.00	0.00	0.00	0.00	0.00	0.00	0.00	0.00	0.00	0.01
	X_5	0.00	0.00	0.00		0.00	0.00	0.00	0.00	0.00	0.00	0.00	0.00	0.00	0.00
	X_6	0.00	0.00	0.00	0.00		0.00	0.00	0.00	0.00	0.00	0.00	0.00	0.00	0.00
	X_7	0.00	0.00	0.00	0.00	0.00		0.00	0.00	0.00	0.00	0.00	0.00	0.00	0.00
	X_9	0.00	0.00	0.00	0.00	0.00	0.00		0.00	0.00	0.00	0.00	0.00	0.00	0.00
	X_{10}	0.00	0.00	0.00	0.00	0.00	0.00	0.00		0.00	0.00	0.00	0.00	0.00	0.00
	X_{12}	0.00	0.00	0.00	0.00	0.00	0.00	0.00	0.00		0.00	0.00	0.00	0.00	0.00
	X_{14}	0.00	0.00	0.00	0.00	0.00	0.00	0.00	0.00	0.00		0.00	0.00	0.00	0.00
	X_{15}	0.00	0.01	0.00	0.00	0.00	0.00	0.00	0.00	0.00	0.00		0.00	0.00	0.00
	X_{19}	0.00	0.00	0.00	0.00	0.00	0.00	0.00	0.00	0.00	0.00	0.00		0.00	0.00
	X_{20}	0.00	0.00	0.00	0.00	0.00	0.00	0.00	0.00	0.00	0.00	0.00	0.00		0.00
	X_{21}	0.00	0.01	0.01	0.00	0.00	0.00	0.00	0.00	0.00	0.00	0.00	0.00	0.00	
	X_{22}	0.00	0.00	0.00	0.00	0.00	0.00	0.00	0.00	0.00	0.00	0.00	0.00	0.00	0.00
	X_{23}	0.00	0.00	0.00	0.00	0.00	0.00	0.00	0.00	0.00	0.00	0.00	0.00	0.00	0.00
	X_{24}	0.00	0.00	0.00	0.00	0.00	0.00	0.00	0.00	0.00	0.00	0.00	0.00	0.00	0.00
	X_{25}	0.00	0.00	0.00	0.00	0.00	0.00	0.00	0.00	0.00	0.00	0.00	0.00	0.00	0.00
	X_{26}	0.00	0.00	0.00	0.00	0.00	0.00	0.00	0.00	0.00	0.00	0.00	0.00	0.00	0.00
	X_{27}	0.00	0.00	0.00	0.00	0.00	0.00	0.00	0.00	0.00	0.00	0.00	0.00	0.00	0.00
	X_{28}	0.00	0.00	0.00	0.00	0.00	0.00	0.00	0.00	0.00	0.00	0.00	0.00	0.00	0.00
	X_{29}	0.00	0.00	0.00	0.00	0.00	0.00	0.00	0.00	0.00	0.00	0.00	0.00	0.00	0.00
	X_{30}	0.00	0.00	0.00	0.00	0.00	0.00	0.00	0.00	0.00	0.00	0.00	0.00	0.00	0.00
	X_{31}	0.00	0.00	0.00	0.00	0.00	0.00	0.00	0.00	0.00	0.00	0.00	0.00	0.00	0.01
	X_{33}	0.00	0.02	0.01	0.00	0.00	0.00	0.00	0.00	0.00	0.00	0.00	0.00	0.00	0.00
	X_{34}	0.00	0.00	0.00	0.00	0.00	0.00	0.00	0.00	0.00	0.00	0.00	0.00	0.00	0.00
	X_{35}	0.00	0.00	0.00	0.00	0.00	0.00	0.00	0.00	0.00	0.00	0.00	0.00	0.00	0.00

(continued)

Table 8.8 (continued)

Variables		X_{22}	X_{23}	X_{24}	X_{25}	X_{26}	X_{27}	X_{28}	X_{29}	X_{30}	X_{31}	X_{33}	X_{34}	X_{35}
Sig. (1-tailed)	X_2	0.00	0.00	0.00	0.00	0.00	0.00	0.00	0.00	0.00	0.00	0.00	0.00	0.00
	X_3	0.00	0.00	0.00	0.00	0.00	0.00	0.00	0.00	0.00	0.00	0.02	0.00	0.00
	X_4	0.00	0.00	0.00	0.00	0.00	0.00	0.00	0.00	0.00	0.00	0.01	0.00	0.00
	X_5	0.00	0.00	0.00	0.00	0.00	0.00	0.00	0.00	0.00	0.00	0.00	0.00	0.00
	X_6	0.00	0.00	0.00	0.00	0.00	0.00	0.00	0.00	0.00	0.00	0.00	0.00	0.00
	X_7	0.00	0.00	0.00	0.00	0.00	0.00	0.00	0.00	0.00	0.00	0.00	0.00	0.00
	X_9	0.00	0.00	0.00	0.00	0.00	0.00	0.00	0.00	0.00	0.00	0.00	0.00	0.00
	X_{10}	0.00	0.00	0.00	0.00	0.00	0.00	0.00	0.00	0.00	0.00	0.00	0.00	0.00
	X_{12}	0.00	0.00	0.00	0.00	0.00	0.00	0.00	0.00	0.00	0.00	0.00	0.00	0.00
	X_{14}	0.00	0.00	0.00	0.00	0.00	0.00	0.00	0.00	0.00	0.00	0.00	0.00	0.00
	X_{15}	0.00	0.00	0.00	0.00	0.00	0.00	0.00	0.00	0.00	0.00	0.00	0.00	0.00
	X_{19}	0.00	0.00	0.00	0.00	0.00	0.00	0.00	0.00	0.00	0.00	0.00	0.00	0.00
	X_{20}	0.00	0.00	0.00	0.00	0.00	0.00	0.00	0.00	0.00	0.00	0.00	0.00	0.00
	X_{21}	0.00	0.00	0.00	0.00	0.00	0.00	0.00	0.00	0.00	0.01	0.00	0.00	0.00
	X_{22}		0.00	0.00	0.00	0.00	0.00	0.00	0.00	0.00	0.00	0.00	0.00	0.00
	X_{23}	0.00		0.00	0.00	0.00	0.00	0.00	0.00	0.00	0.00	0.00	0.00	0.00
	X_{24}	0.00	0.00		0.00	0.00	0.00	0.00	0.00	0.00	0.00	0.00	0.00	0.00
	X_{25}	0.00	0.00	0.00		0.00	0.00	0.00	0.00	0.00	0.00	0.00	0.00	0.00
	X_{26}	0.00	0.00	0.00	0.00		0.00	0.00	0.00	0.00	0.00	0.00	0.00	0.00
	X_{27}	0.00	0.00	0.00	0.00	0.00		0.00	0.00	0.00	0.00	0.00	0.00	0.00
	X_{28}	0.00	0.00	0.00	0.00	0.00	0.00		0.00	0.00	0.00	0.00	0.00	0.00
	X_{29}	0.00	0.00	0.00	0.00	0.00	0.00	0.00		0.00	0.00	0.00	0.00	0.00
	X_{30}	0.00	0.00	0.00	0.00	0.00	0.00	0.00	0.00		0.00	0.00	0.00	0.00
	X_{31}	0.00	0.00	0.00	0.00	0.00	0.00	0.00	0.00	0.00		0.00	0.00	0.00
	X_{33}	0.00	0.00	0.00	0.00	0.00	0.00	0.00	0.00	0.00	0.00		0.00	0.00
	X_{34}	0.00	0.00	0.00	0.00	0.00	0.00	0.00	0.00	0.00	0.00	0.00		0.00
	X_{35}	0.00	0.00	0.00	0.00	0.00	0.00	0.00	0.00	0.00	0.00	0.00	0.00	

Table 8.9 Anti-image correlation

Var.	X_2	X_3	X_4	X_5	X_6	X_7	X_9	X_{10}	X_{12}	X_{14}	X_{15}	X_{19}	X_{20}	X_{21}
X_2	0.928[a]	−0.155	−0.178	−0.073	0.062	−0.161	0.128	−0.090	−0.245	0.029	0.124	−0.048	0.100	−0.015
X_3	−0.155	0.897[a]	−0.133	−0.121	−0.082	−0.094	0.000	−0.014	0.040	0.194	0.100	0.120	0.167	−0.071
X_4	−0.178	−0.133	0.846[a]	−0.155	0.117	0.024	−0.285	0.002	−0.088	−0.153	−0.084	0.162	−0.094	0.205
X_5	−0.073	−0.121	−0.155	0.928[a]	−0.089	−0.049	−0.055	0.040	−0.027	−0.045	−0.039	−0.338	0.121	0.047
X_6	0.062	−0.082	0.117	−0.089	0.945[a]	−0.432	−0.099	−0.050	−0.095	−0.212	0.024	0.029	−0.102	0.048
X_7	−0.161	−0.094	0.024	−0.049	−0.432	0.933[a]	−0.110	0.046	−0.029	−0.024	−0.121	0.008	−0.086	0.117
X_9	0.128	0.000	−0.285	−0.055	−0.099	−0.110	0.915[a]	−0.171	−0.124	−0.060	0.175	−0.188	0.202	−0.092
X_{10}	−0.090	−0.014	0.002	0.040	−0.050	0.046	−0.171	0.934[a]	0.068	−0.045	−0.122	0.064	0.112	−0.042
X_{12}	−0.245	0.040	−0.088	−0.027	−0.095	−0.029	−0.124	0.068	0.949[a]	−0.176	−0.168	0.005	−0.008	−0.171
X_{14}	0.029	0.194	−0.153	−0.045	−0.212	−0.024	−0.060	−0.045	−0.176	0.925[a]	−0.182	0.020	−0.061	−0.037
X_{15}	0.124	0.100	−0.084	−0.039	0.024	−0.121	0.175	−0.122	−0.168	−0.182	0.935[a]	−0.123	0.035	0.067
X_{19}	−0.048	0.120	0.162	−0.338	0.029	0.008	−0.188	0.064	0.005	0.020	−0.123	0.925[a]	−0.103	−0.099
X_{20}	0.100	0.167	−0.094	0.121	−0.102	−0.086	0.202	0.112	−0.008	−0.061	0.035	−0.103	0.910[a]	−0.474
X_{21}	−0.015	−0.071	0.205	0.047	0.048	0.117	−0.092	−0.042	−0.171	−0.037	0.067	−0.099	−0.474	0.879[a]
X_{22}	−0.067	−0.113	−0.043	0.116	0.029	−0.099	−0.178	−0.206	−0.059	−0.086	−0.011	−0.061	−0.313	−0.002
X_{23}	0.061	−0.273	0.130	−0.248	0.035	0.160	0.028	−0.172	−0.062	−0.191	−0.133	0.053	−0.184	−0.063
X_{24}	−0.001	0.120	0.037	−0.092	0.028	−0.114	−0.055	−0.259	0.035	−0.011	0.010	0.027	−0.066	0.012
X_{25}	0.115	−0.256	−0.002	0.083	−0.014	0.004	0.063	0.206	0.074	−0.201	0.001	−0.046	0.117	−0.053
X_{26}	−0.035	0.023	−0.291	0.062	−0.051	−0.085	0.085	−0.007	0.218	−0.083	0.107	−0.048	0.085	−0.133
X_{27}	−0.113	−0.033	−0.311	0.168	−0.123	0.010	0.106	0.041	0.050	0.150	0.041	−0.192	0.008	−0.225
X_{28}	0.100	−0.134	−0.027	−0.049	−0.008	0.002	0.185	−0.018	−0.081	−0.172	−0.038	0.022	−0.183	0.147
X_{29}	0.119	0.017	−0.292	0.189	−0.027	−0.131	0.184	−0.086	−0.015	−0.115	0.135	−0.151	0.062	−0.055
X_{30}	−0.117	0.098	0.166	−0.149	0.071	−0.131	−0.168	0.001	0.032	0.181	−0.082	0.208	−0.227	−0.001
X_{31}	−0.079	−0.071	0.352	−0.177	0.079	0.194	−0.223	0.010	−0.146	−0.085	0.040	−0.071	−0.028	0.312
X_{33}	−0.202	0.032	0.199	0.024	−0.026	0.061	−0.115	0.001	0.042	−0.268	−0.083	−0.032	−0.127	−0.002
X_{34}	−0.035	−0.080	−0.064	−0.203	−0.008	0.078	−0.129	−0.164	−0.190	0.141	−0.042	−0.039	−0.049	0.028
X_{35}	−0.077	0.075	−0.053	0.048	0.031	−0.137	−0.166	0.089	0.145	0.234	−0.345	0.094	−0.169	−0.082

(continued)

Table 8.9 (continued)

Var.	X_{22}	X_{23}	X_{24}	X_{25}	X_{26}	X_{27}	X_{28}	X_{29}	X_{30}	X_{31}	X_{33}	X_{34}	X_{35}
X_2	−0.067	0.061	−0.001	0.115	−0.035	−0.113	0.100	0.119	−0.117	−0.079	−0.202	−0.035	−0.077
X_3	−0.113	−0.273	0.120	−0.256	0.023	−0.033	−0.134	0.017	0.098	−0.071	0.032	−0.080	0.075
X_4	−0.043	0.130	0.037	−0.002	−0.291	−0.311	−0.027	−0.292	0.166	0.352	0.199	−0.064	−0.053
X_5	0.116	−0.248	−0.092	0.083	0.062	0.168	−0.049	0.189	−0.149	−0.177	0.024	−0.203	0.048
X_6	0.029	0.035	0.028	−0.014	−0.051	−0.123	−0.008	−0.027	0.071	0.079	−0.026	−0.008	0.031
X_7	−0.099	0.160	−0.114	0.004	−0.085	0.010	0.002	−0.131	−0.131	0.194	0.061	0.078	−0.137
X_9	−0.178	0.028	−0.055	0.063	0.085	0.106	0.185	0.184	−0.168	−0.223	−0.115	−0.129	−0.166
X_{10}	−0.206	−0.172	−0.259	0.206	−0.007	0.041	−0.018	−0.086	0.001	0.010	0.001	−0.164	0.089
X_{12}	−0.059	−0.062	0.035	0.074	0.218	0.050	−0.081	−0.015	0.032	−0.146	0.042	−0.190	0.145
X_{14}	−0.086	−0.191	−0.011	−0.201	−0.083	0.150	−0.172	−0.115	0.181	−0.085	−0.268	0.141	0.234
X_{15}	−0.011	−0.133	0.010	0.001	0.107	0.041	−0.038	0.135	−0.082	0.040	−0.083	−0.042	−0.345
X_{19}	−0.061	0.053	0.027	−0.046	−0.048	−0.192	0.022	−0.151	0.208	−0.071	−0.032	−0.039	0.094
X_{20}	−0.313	−0.184	−0.066	0.117	0.085	0.008	−0.183	0.062	−0.227	−0.028	−0.127	−0.049	−0.169
X_{21}	−0.002	−0.063	0.012	−0.053	−0.133	−0.225	0.147	−0.055	−0.001	0.312	−0.002	0.028	−0.082
X_{22}	0.939[a]	−0.083	0.165	−0.231	−0.031	0.059	0.122	0.126	0.092	−0.193	0.017	−0.056	0.125
X_{23}	−0.083	0.923[a]	−0.063	0.128	−0.224	−0.107	0.002	−0.071	0.084	0.032	0.085	0.141	−0.339
X_{24}	0.165	−0.063	0.884[a]	−0.482	−0.153	−0.171	−0.001	0.032	−0.037	−0.010	−0.114	0.263	0.072
X_{25}	−0.231	0.128	−0.482	0.893[a]	−0.178	0.062	0.088	0.066	−0.218	−0.047	−0.066	−0.209	−0.122
X_{26}	−0.031	−0.224	−0.153	−0.178	0.922[a]	0.119	−0.009	0.137	−0.048	−0.107	−0.025	−0.162	0.057
X_{27}	0.059	−0.107	−0.171	0.062	0.119	0.901[a]	−0.236	0.127	−0.136	−0.267	0.082	−0.234	0.099
X_{28}	0.122	0.002	−0.001	0.088	−0.009	−0.236	0.937[a]	0.035	−0.234	−0.129	−0.005	0.041	−0.223
X_{29}	0.126	−0.071	0.032	0.066	0.137	0.127	0.035	0.861[a]	−0.426	−0.379	0.013	−0.111	−0.078
X_{30}	0.092	0.084	−0.037	−0.218	−0.048	−0.136	−0.234	−0.426	0.909[a]	−0.100	−0.012	−0.070	0.101
X_{31}	−0.193	0.032	−0.010	−0.047	−0.107	−0.267	−0.129	−0.379	−0.100	0.883[a]	0.039	0.116	−0.062
X_{33}	0.017	0.085	−0.114	−0.066	−0.025	0.082	−0.005	0.013	−0.012	0.039	0.949[a]	−0.082	−0.109
X_{34}	−0.056	0.141	0.263	−0.209	−0.162	−0.234	0.041	−0.111	−0.070	0.116	−0.082	0.939[a]	−0.196
X_{35}	0.125	−0.339	0.072	−0.122	0.057	0.099	−0.223	−0.078	0.101	−0.062	−0.109	−0.196	0.910[a]

[a] The measure of sampling adequacy *MSA*

Table 8.10 Component analysis

Factor	Initial eigenvalues		
	Total	% of variance	Cumulative %
1	*12.595*	*46.649*	*46.649*
2	*1.626*	*6.022*	*52.670*
3	*1.490*	*5.517*	*58.187*
4	*1.227*	*4.544*	*62.731*
5	*1.046*	*3.875*	*66.607*
6	0.900	3.333	69.940
7	0.803	2.975	72.915
8	0.709	2.625	75.539
9	0.639	2.365	77.905
10	0.633	2.344	80.248
11	0.610	2.258	82.507
12	0.559	2.069	84.576
13	0.469	1.736	86.312
14	0.424	1.572	87.883
15	0.412	1.526	89.410
16	0.391	1.448	90.858
17	0.367	1.359	92.216
18	0.326	1.207	93.424
19	0.283	1.048	94.472
20	0.267	0.990	95.461
21	0.237	0.878	96.339
22	0.207	0.765	97.104
23	0.205	0.760	97.865
24	0.171	0.634	98.499
25	0.162	0.601	99.100
26	0.131	0.486	99.586
27	0.112	0.414	100.000

Italics are those the factors having eigenvalues greater than 1

with eigenvalues greater than 1. The five factors represent about 67 % of the variance of the 27 variables, meeting the requirement of total variance explained. As a result, five factors are retained for the computation of factor matrix of loadings. Results in Table 8.11 present the unrotated component analysis factor matrix for the five factors that are extracted. The numbers shown in the first five columns of the Table are factor loadings of each variable on each of the factors. Given the sample size of 126, factor loadings of 0.5 or higher will be considered significant for factor interpretation (Hair et al. 2010). As shown in Table 8.11, all variables have high loadings. However, factor interpretation would be difficult since all variables have high loadings only on the first factor, and one variable (X_{24}) has cross-loadings. Hence, factor rotation should be used to improve the interpretation by resulting in more meaningful factor pattern (Hair et al. 2010). Table 8.11 also provides the

Table 8.11 Factor matrix

	Factor					Communality
	1	2	3	4	5	
X_2	0.612	0.394	0.020	−0.146	0.036	0.553
X_3	0.537	0.460	0.124	0.006	0.364	0.648
X_4	0.606	0.423	0.088	−0.110	0.260	0.634
X_5	0.718	0.264	−0.007	−0.122	−0.156	0.625
X_6	0.715	−0.038	0.168	−0.026	−0.155	0.566
X_7	0.731	0.024	0.102	0.067	−0.086	0.557
X_9	0.721	0.269	0.202	−0.117	−0.238	0.703
X_{10}	0.626	0.093	0.216	−0.092	−0.094	0.464
X_{12}	0.744	0.081	−0.084	−0.327	−0.188	0.710
X_{14}	0.757	−0.223	0.231	−0.031	−0.243	0.736
X_{15}	0.676	−0.310	−0.059	−0.249	−0.054	0.622
X_{19}	0.632	0.017	−0.047	−0.176	−0.263	0.502
X_{20}	0.760	−0.442	−0.137	−0.094	0.156	0.826
X_{21}	0.585	−0.430	0.036	−0.197	0.321	0.671
X_{22}	0.754	0.009	0.260	−0.140	−0.028	0.657
X_{23}	0.764	−0.213	0.008	−0.165	0.199	0.695
X_{24}	0.585	−0.174	0.297	0.536	−0.024	0.747
X_{25}	0.692	−0.039	0.298	0.443	0.073	0.770
X_{26}	0.629	0.035	0.383	0.287	0.264	0.696
X_{27}	0.676	0.156	−0.330	0.075	0.271	0.669
X_{28}	0.685	−0.143	−0.437	0.108	0.109	0.704
X_{29}	0.597	0.091	−0.479	0.261	−0.166	0.689
X_{30}	0.723	0.002	−0.381	0.351	−0.037	0.793
X_{31}	0.679	0.185	−0.324	0.220	−0.299	0.739
X_{33}	0.640	−0.330	0.193	0.034	−0.228	0.609
X_{34}	0.787	0.235	−0.076	−0.116	0.117	0.707
X_{35}	0.726	−0.251	−0.198	−0.132	0.211	0.691

communalities of the variables. Communality of each variable reflects the amount of variance accounted for by the factor solution for each variable. For variables to meet the acceptance levels of explanation, at least one-half of the variance of each variable in this study should be taken into account. In other words, communality of each variable should be at least 0.5. There is only variable X_{10} having communality lower than 0.5. Variable X_{10} is thus removed from the rotation of the factor matrix.

VARIMAX was used to rotate factors and estimate rotated factor loadings for each variable. As shown in Table 8.12, the rotated factor solution improves the structure considerably. There are no variables cross-loading on two factors. All variables have sufficient communality, which is more than 0.5. Moreover, all of the loadings above 0.5 are redistributed among factors rather than gathering in one single factor. Variable X_7 does not have high loadings (loadings less than 0.5), thus

Table 8.12 Rotated factor-loading matrix

Rotated factor matrix						Communality
	Factor					
	1	2	3	4	5	
X_2	0.397	0.096	0.209	0.585	0.066	0.556
X_3	0.089	0.094	0.116	0.759	0.211	0.651
X_4	0.232	0.153	0.136	0.722	0.135	0.636
X_5	0.564	0.145	0.302	0.432	0.122	0.632
X6	0.551	0.281	0.169	0.218	0.344	0.577
X_7	0.467	0.253	0.260	0.282	0.373	0.568
X_9	0.658	0.081	0.168	0.415	0.237	0.695
X_{12}	0.649	0.347	0.268	0.320	−0.024	0.716
X_{14}	0.623	0.384	0.139	0.066	0.422	0.736
X_{15}	0.450	0.604	0.195	0.068	0.099	0.620
X_{19}	0.597	0.240	0.258	0.168	0.071	0.515
X_{20}	0.270	0.775	0.298	0.084	0.235	0.824
X_{21}	0.139	0.768	0.022	0.141	0.199	0.669
X_{22}	0.528	0.360	0.061	0.362	0.327	0.650
X_{23}	0.292	0.659	0.192	0.298	0.224	0.695
X_{24}	0.161	0.158	0.239	0.050	0.796	0.745
X_{25}	0.227	0.186	0.221	0.265	0.757	0.777
X_{26}	0.127	0.239	0.047	0.424	0.665	0.698
X_{27}	0.072	0.358	0.536	0.484	0.108	0.667
X_{28}	0.128	0.494	0.633	0.173	0.122	0.705
X_{29}	0.214	0.132	0.777	0.128	0.112	0.696
X_{30}	0.182	0.258	0.753	0.182	0.304	0.792
X_{31}	0.416	0.043	0.705	0.188	0.177	0.739
X_{33}	0.518	0.384	0.128	−0.067	0.423	0.615
X_{34}	0.379	0.325	0.339	0.570	0.133	0.707
X_{35}	0.221	0.680	0.329	0.234	0.127	0.691

is not considered in the interpretation of factors. Rearranging the factor loadings, the pattern emerges (see Table 8.13).

As shown in Table 8.13, with VARIMAX rotation, the factor structure for the remaining 25 variables is now defined, comprising five distinct groups of variables, which can be utilized in further data analysis. Factor 1 is heavily loaded on variables:

- Lack of inter-ministerial and intersectoral coordination (X_{12});
- Political leaders and government bureaucrats make decisions in their personal, sectoral, or regional interests (for example securing political positions or competing for scarce funds) as opposed to national goals (X_6);

Table 8.13 Rearrangement of rotated factor matrix

Rotated factor matrix[a]					
	Factor				
	1	2	3	4	5
X_{12}	0.649				
X_9	0.658				
X_{14}	0.623				
X_{19}	0.597				
X_5	0.564				
X_6	0.551				
X_{22}	0.528				
X_{33}	0.518				
X_{20}		0.775			
X_{21}		0.768			
X_{35}		0.680			
X_{23}		0.659			
X_{15}		0.604			
X_{29}			0.777		
X_{30}			0.753		
X_{31}			0.705		
X_{28}			0.633		
X_{27}			0.536		
X_3				0.759	
X_4				0.722	
X_2				0.585	
X_{34}				0.570	
X_{24}					0.796
X_{25}					0.757
X_{26}					0.665

[a] Loading less than 0.5 are not shown and variables are sorted by highest loading

- Incompetent and unqualified civil servants in infrastructure planning agencies (X_{14});
- Lack of independent peer reviews and quality checks on forecasts and planning outcomes by independent-review bodies and scientific and professional community (X_9);
- Lack of a high-powered government institution that provides an effective mechanism for implementing national infrastructure plans (X_{19});
- Inadequacy in ex-post analysis on whether projects meet objectives (X_5);
- Inadequate financial and physical auditing for infrastructure construction (X_{22}); and
- Changing and inconsistent law and regulations in infrastructure construction (X_{33}).

These variables centre on issues relating to the government's capacity in planning, coordinating and monitoring infrastructure development, and commitment of political leaders. Factor 1, therefore, can be called insufficient institutional capacity of the government.

Factor 2 is characterized by variables:

- Lack of transparency and competitiveness in bid processes (X_{20});
- Discretionary power of individual bureaucrats involved in award of contracts (X_{21});
- Bureaucracy in formal procedures relating to infrastructure project planning, construction permission, and administration (X_{35});
- Inadequate capacity of regulatory bodies to enforce regulations of corruption (X_{23}); and
- Complicated and bureaucratic administrative system in infrastructure planning agencies (X_{15}).

These variables describe weaknesses in transparency and accountability in infrastructure development. Factor 2 can thus be called lack of transparency and accountability in infrastructure development.

Factor 3 represents the group of five variables, including:

- Lack of a land acquisition compensation-monitoring system (X_{29});
- Lack of clarity about valuation methods for compensation (X_{30});
- Lack of law enforcement to regulate the price of land acquired for infrastructure projects (X_{31});
- Bureaucracy in settling land disputes and claims (X_{28}); and
- Lack of a negotiating mechanism to make land acquisition compensation more market-oriented (X_{27}).

Straightforwardly, Factor 3 can be named as lack of an effective land acquisition framework.

Factor 4 is defined by variables:

- Inherent problems in predicting the future and monetizing external and indirect effects (X_3);
- Lack of experienced forecasters (X_4);
- Inadequate data (X_2); and
- Ineffectiveness of implementation of existing statutes and codes in infrastructure construction (X_{34}).

Since X_{34} has the lowest loading among these variables, and all variables X_2, X_3 and X_4 refers to issues related to the capacity of forecasting. Factor 4 is named as inadequate forecasting capacity.

Factor 5 is highly loaded on variables:

- Poor level of efficiency and quality of work (X_{24});
- Poor level of professionalism and entrepreneurship (X_{25}); and

- Resources shortages, especially in construction technology, management, and finance (X_{26}).

Factor 5, thus, is called insufficient building capacity of local firms.

8.3.3.3 Factor Interpretation

Factor 1 *Insufficient institutional capacity of the government*

This factor raises questions about the institutional capacity and political leaders' commitment. The key question of such capacity lies in the ability of the government to create a strong planning authority for spearheading infrastructure development. Strong government leadership and commitment are considered crucial in determining infrastructure development strategies (Mody 1997). However, several variables (X_6, X_9, X_{12}, X_{14}, X_{19}) included in this factor suggests that there is a lack of commitment to national goals and planning agencies in Vietnam are fragmented. As a result, they fail to serve as focal points for inter-ministerial and inter-agency coordination, and as vehicles that enable political leaders and government bureaucrats to develop collaboration in infrastructure development. The insufficient institutional capacity also restricts planning agencies to determine priorities and organize resources with a broader vision. Other variables of Factor 1 emphasize on the limit of monitoring aspect of institutional capacity. Two constraints in this aspect are highlighted, including lack of ex-post analysis on whether projects meet objectives (X_5), and insufficient financial and physical auditing for infrastructure construction (X_{22}). With these constraints, planning agencies are prevented from being well informed of how the plans are implemented and how the resources are allocated. Planning outcomes are difficult to achieve without ongoing assessment of the effectiveness of plan implementation. The deficiencies in government capacities become more critical in such a developing country as Vietnam where the government still account for a majority of funding invested in infrastructure construction. In addition, changing and inconsistent law and regulations in infrastructure construction (X_{33}) also reflects the weaknesses of the institutions in supporting efficient infrastructure delivery.

Factor 2 *Lack of transparency and accountability in infrastructure development*

Factor 2 directs attention to areas associated with transparency and accountability. Variables with high loadings on Factor 2 include the lack of transparency and competitiveness in bid processes (X_{20}), the abuse of power by the bureaucrats involved in contract awarding (X_{21}) and the weak enforcement abilities of regulatory bodies (X_{23}). The survey evidence also classifies bureaucracy in administrative system and formal procedures relating to infrastructure construction into this factor group (X_{15} and X_{35}). The occurrence of these problems highly correlates with each other, implying the low level of transparency and accountability in funded infrastructure construction in Vietnam. The variables highlighted in this factor suggest that the lack of accountability can be found in all stages of infrastructure construction from government planning, contract awarding to the execution of physical facilities. The consequences of this factor would range from the low efficiency of services delivered by government agencies to the increasing bureaucratic corruption in infrastructure construction. Corruption in infrastructure construction is found to be significant (Kenny 2007). More importantly, it would affect the quality of infrastructure construction and ultimately on the delivery of infrastructure services. The lack of transparency and accountability thus can have significantly negative impact on the efficiency of funded infrastructure construction.

Factor 3 *Lack of an effective land acquisition framework*

The factor group describes problems with the current framework of land acquisition for infrastructure construction in Vietnam. The survey evidence identifies that land acquisition problems are among those having the highest degree of occurrence. Two major aspects contributing to the factor include land prices and regulations. With land prices, the attention focuses on approach to valuation, and methods of valuation and monitoring (X_{27}, X_{30}, X_{29}). It is suggested that the current approach to valuation does not adequately reflect the market price of land. Methods of valuation are not clearly defined, and a monitoring system is not in place to ensure that the land owner gets the correct value for the land. The factor also raises concerns about the regulatory framework used to control land price and the process of acquisition (X_{28}, X_{31}). Increasing land demand for infrastructure would boost land price to an unreasonably high level, thus requiring government to have a control policy. As a result, lack of tools for government to improve market efficiency for land acquired for infrastructure could make it hard to meet land demand driven by rapid growth. Finally, bureaucracy in setting land disputes and claims is listed in this factor. The presence of the issue implies that the current land management practices could pose significant risks to the success of the land acquisition process.

Factor 4 *Inadequate forecasting capacity*

This factor presents deficiencies in forecasting for government planning. Attention is directed to inherent problems in forecasting and concerns about input data and human resources (X_2, X_3, X_4). The existence of inherent problems in predicting the future and monetizing external and indirect effects reduces forecast accuracy. As Vietnam is open to international trade, aid, and foreign investments, the country is more vulnerable to external economic disturbances. Unanticipated external effects can thus induce serious errors in forecasting. The statistical results also placed inadequate data and lack of experienced forecasters in this factor group. This reveals problems in the quality of statistical data and planning personnel used in the formulation of development plans in Vietnam. Interestingly, X_{34} is also included in this factor. Collectively, these problems diminish the government's forecasting ability and subsequently the value of development plans.

Factor 5 *Insufficient building capacity of local firms*

Variables highly loading on this factor emphasize on issues with local construction firms in Vietnam. The statistical results support the view that local construction firm problems can be reflected by problems of construction quality, problems of professionalism, and problems of resource shortages in construction technology, management, and finance (X_{24}, X_{25}, X_{26}). Taken together these problems enhance the understanding of building capacity of local construction firms in meeting growing demand for infrastructure construction.

8.4 Structural Equation Modelling (SEM)

8.4.1 Hypotheses and the Model

In Sect. 8.2, the study discovered that the efficiency of public investments in infrastructure construction to support trade and economic growth in Vietnam is falling. The objective of this stage of data analysis is to capture the relationship between causal factors and the outcomes of the inefficient use of public investments in infrastructure. As mentioned in Chap. 7, the unexpected outcomes refer to the occurrence of the following:

- Poorly selected infrastructure projects (IU_1);
- Poorly executed and ineffective infrastructure projects (IU_2); and
- Waste of national resources in funded infrastructure projects (IU_3).

Results of the survey have shown that IU_3 (waste of national resources) and IU_2 (poorly executed and ineffective infrastructure projects) were rated as having moderate-to-high occurrence. The considerable occurrence of IU_1 (poorly selected infrastructure projects) was also observed (Table 8.14).

From the exploratory factor analysis, five key factors could be extracted from variables that might be related to the outcomes of the inefficient use of public investments in infrastructure. Their conceptual definitions and reflective indicators are as follows:

- Insufficient institutional capacity of the government (IC): Institutional capacity and commitment in this context implies the strength of political leaders and government authority for spearheading infrastructure development. Reflective indicator variables are X_5, X_6, X_9, X_{12}, X_{14}, X_{19}, X_{22}, and X_{33}.
- Lack of transparency and accountability in infrastructure construction (TA): Transparency and accountability in infrastructure construction refer to the government being open and accountable to the bureaucratic system in contract awarding, project planning, construction permission, and administration. Reflective indicator variables are X_{15}, X_{20}, X_{21}, X_{23}, and X_{35}.

Table 8.14 Ranking of the outcomes of the inefficient use of public investments in infrastructure

Rank	Outcomes of the inefficient use of public investments	Median	SD
1	Waste of national resources in funded infrastructure projects	3.4	1.1
2	Poorly executed and ineffective infrastructure construction	3.1	0.87
3	Poorly selected infrastructure projects	2.4	0.89

- Lack of an effective land acquisition framework (LA): Land acquisition framework in this study concerns the environment that is created to facilitate land acquisition for infrastructure construction. Reflective indicator variables are X_{27}, X_{28}, X_{29}, X_{30}, and X_{31}.
- Inadequate forecasting capacity (FC): Forecasting capacity refers to government's ability to formulate quantitative plans for infrastructure development. Reflective indicator variables are X_2, X_3, X_4, and X_{34}.
- Insufficient building capacity of local firms (BC): Supply capacity of local construction firms refers to the ability of local construction firms to meet demand for infrastructure construction. Reflective indicator variable are X_{24}, X_{25}, and X_{26}.

From the literature review, these factors are hypothesized to affect the efficient use of public investments in infrastructure in different ways. Insufficient institutional capacity of the government is hypothesized to be the core factor as it would have critical effects on other factors, which in turn affects investment efficiency of funded infrastructure development. More specifically, these hypotheses are:

H_1 Insufficient institutional capacity of the government is positively related to inadequate forecasting capacity for infrastructure development (IC +→ FC)
H_2 Insufficient institutional capacity of the government is positively related to lack of transparency and accountability in infrastructure construction (IC +→ TA)
H_3 Insufficient institutional capacity of the government is positively related to lack of an effective land acquisition framework (IC +→ LA)
H_4 Insufficient institutional capacity of the government is positively related to insufficient supply capacity of local construction firms in meeting demand for infrastructure development (IC +→ BC)
H_5 Inadequate forecasting capacity is positively related to the inefficient use of public investments in infrastructure (FC +→ IU)
H_6 Lack of transparency and accountability in infrastructure development is positively related to the inefficient use of public investments in infrastructure (TA +→ IU)
H_7 Lack of an effective land acquisition framework is positively related to the inefficient use of public investments in infrastructure (LA +→ IU)
H_8 Insufficient building capacity of local firms is positively related to the inefficient use of public investments in infrastructure (BC +→ IU)

These hypotheses are visually illustrated with Fig. 8.2. As shown in Fig. 8.2, the model displays 28 measured indicator variables and six latent constructs (IU, IC, TA, LA, FC, and BC). While measured variables are shown with a box, latent constructs are shown with an oval. IC is exogenous (independent) construct in this model as there is no single-headed arrow entering the construct. IC is indicated by eight measured variables. TA, LA, BC, LC, and IU are all endogenous (dependent) constructs. TA, LA, BC, and LC are all hypothesized as outcomes in some relationships and as predictors in others. TA and LA are indicated by five measured items; FC is indicated by four measured items; and BC is indicated by three measured variables. IU is the final outcome and is indicated by three measured variables. Hypotheses to be tested are represented by arrows connecting two constructs and parameter estimates *P*. For example, H_1 linking FC-IU is represented by $P_{IU,FC}$.

As discussed in Sect. 7.5.2, between CB-SEM and PLS-SEM, the latter is more appropriate to examine the multiple interrelated dependence relationships in this study given its prediction orientation. Similar to CB-SEM, PLS-SEM assessment of the theoretical model as specified in Fig. 8.2 consists of two sections. The first section, evaluation of measurement model, engages the evaluation of the reliability and validity of measured indicators representing latent constructs. The second section, evaluation of structural model, involves the assessment of parameter estimates representing the structural relationships among latent constructs.

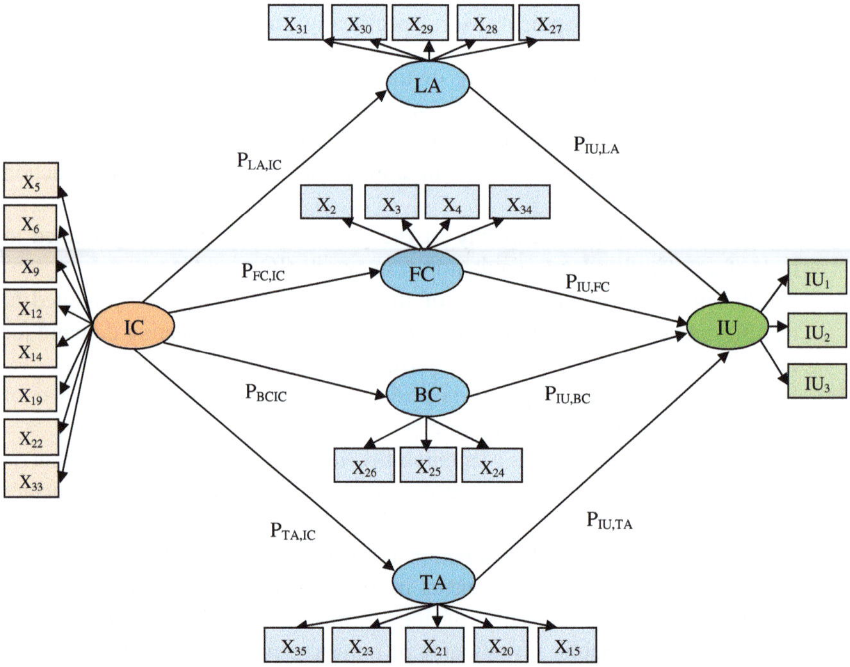

Legend:

IU : Inefficient use of public investments in infrastructure
IC : Insufficient institutional capacity of the government
TA : Lack of transparency and accountability in infrastructure development
LA : Lack of an effective land acquisition framework
FC : Inadequate forecasting capacity
BC : Insufficient building capacity of local firms
P : Parameter estimates

Fig. 8.2 Model of the inefficient use of public investments in infrastructure in Vietnam

8.4.2 Evaluation of the Measurement Model

As mentioned earlier, the purpose of the evaluation is to assess the reliability and validity of measurement. This step is important as it provides the evidence of the extent to which the measured indicators actually represent the theoretical latent constructs. The evaluation focuses on construct reliability, convergent validity, and discriminant validity.

Construct reliability considers two elements: internal consistency reliability and indicator reliability. A construct's internal consistency reliability reflects the consistency that all measured indicators represent the same latent construct. A construct's internal consistency reliability is estimated by a composite reliability value. A composite reliability value should be higher than 0.7 to be regarded as satisfactory (Hair et al. 2011). Table 8.15 shows the composite reliability values of constructs computed using SmartPLS software. All composite reliability values are from 0.8 to 0.9, indicating that internal consistency exists in all latent constructs.

Indicator reliability focuses on loadings on a factor. Each indicator's absolute standardized loading should be higher than 0.7. High-standardized loadings means that measured indicators converge on a common point. As shown in Table 8.15, only two estimates (IU_1 and X_{33}) fall below the 0.7 cut-off. However, loadings of IU_1 and X_{33} can be considered adequate as they are only slightly below 0.7. As a result, IU_1 and X_{33} are retained to support content validity (Hair et al. 2010).

Convergent validity is a summary indicator of convergence. It reflects the extent to which measured indicators of a specific construct converge or share a high proportion of variance in common. The convergence validity is assessed by examining average variance extracted (AVE). AVE is computed using standardized loadings (Hair et al. 2010):

$$\mathrm{AVE} = \frac{\sum_{i=1}^{n} L_i^2}{n}$$

where L_i is standardized factor loading and i is the number of items. A sufficient degree of convergent validity requires an AVE value of 0.5 or higher. That is because an AVE value of 0.5 or higher means that the latent construct explains more than half of its indicators' variance. The AVE values for all six constructs of this study are higher than 0.5 (Table 8.15), suggesting an adequate convergence.

Finally, discriminant validity involves the assessment of the distinction between constructs. In order to ascertain that a construct is unique and captures some phenomena other measures do not. Hair et al. (2011) mentioned two criteria: the Fornell–Larker criterion and cross-loadings. The Fornell–Larker criterion suggests that a latent construct is distinct from others if the construct shares more variance with its assigned indicators than with another latent construct in the structural model. Statistically, the AVE value of each latent construct should be higher than the construct's highest squared correlation with any other latent construct, if discriminant validity exists. Table 8.16 shows the squared correlation between

Table 8.15 Measured indicator loadings

Measured indicators	FC	IC	IU	LA	BC	TA
X_2	0.7969					
X_3	0.7203					
X_4	0.7805					
X_{34}	0.8572					
X_5		0.7511				
X_6		0.7572				
X_9		0.7913				
X_{12}		0.7923				
X_{14}		0.8165				
X_{19}		0.6999				
X_{22}		0.7909				
X_{33}		0.6859				
IU_1			0.6716			
IU_2			0.9285			
IU_3			0.9139			
X_{27}				0.7817		
X_{28}				0.8095		
X_{29}				0.7794		
X_{30}				0.8667		
X_{31}				0.8245		
X_{24}					0.8566	
X_{25}					0.8932	
X_{26}					0.8198	
X_{15}						0.7793
X_{20}						0.8854
X_{21}						0.7554
X_{23}						0.8546
X_{35}						0.8469
Composite reliability	0.8689	0.9169	0.8813	0.9068	0.8923	0.9144
AVE	0.6245	0.5805	0.7161	0.6609	0.7345	0.6819

constructs of this study. As shown in Table 8.16, AVE values of a construct are all higher than the construct's highest squared correlation with other latent constructs (the numbers in bold type). Besides, comparing an indicator's factor loadings with its cross-loadings can also reveal discriminant validity. As shown in Table 8.17, all indicators' loading with its associated latent constructs (the numbers in italic) are higher than their cross-loadings with all the remaining constructs, implying that discriminant validity of constructs of this study is supported.

Table 8.16 Squared correlations of constructs

Squared correlation	FC	IC	IU	LA	BC	TA
FC	1					
IC	**0.5345**	1				
IU	0.2688	0.4143	1			
LA	0.4171	0.4587	0.3232	1		
BC	0.2999	0.4169	0.2766	0.2946	1	
TA	0.3312	**0.5616**	**0.3486**	**0.4328**	**0.3124**	1
AVE	**0.6245**	**0.5805**	**0.7161**	**0.6609**	**0.7345**	**0.6819**

Bolds are the construct's highest squared correlation with other latent constructs and AVE values of a construct (the last row)

Table 8.17 Indicators' cross-loadings

Indicators	FC	IC	IU	LA	BC	TA
X_2	*0.7969*	0.5679	0.4777	0.4823	0.371	0.3937
X_3	*0.7203*	0.4531	0.2901	0.4098	0.4154	0.3492
X_4	*0.7805*	0.5339	0.2878	0.4505	0.4285	0.4077
X_{34}	*0.8572*	0.7091	0.521	0.6511	0.5109	0.6162
X_5	0.6525	*0.7511*	0.4338	0.5685	0.4635	0.525
X_6	0.5262	*0.7572*	0.5218	0.5069	0.5139	0.5607
X_9	0.6459	*0.7913*	0.426	0.507	0.4973	0.4847
X_{12}	0.6427	*0.7923*	0.4659	0.5874	0.3675	0.6132
X_{14}	0.4818	*0.8165*	0.5036	0.5186	0.6039	0.6532
X_{19}	0.463	*0.6999*	0.4995	0.5028	0.3943	0.5126
X_{22}	0.6136	*0.7909*	0.5313	0.4993	0.5615	0.6295
X_{33}	0.4054	*0.6859*	0.5547	0.432	0.5269	0.5809
IU_1	0.3281	0.4197	*0.6716*	0.2361	0.3387	0.238
IU_2	0.4502	0.5571	*0.9285*	0.562	0.523	0.5412
IU_3	0.5139	0.6335	*0.9139*	0.5597	0.4546	0.6258
X_{27}	0.6299	0.5286	0.5126	*0.7817*	0.4386	0.5456
X_{28}	0.4821	0.5407	0.4798	*0.8095*	0.4135	0.6445
X_{29}	0.4576	0.4728	0.3507	*0.7794*	0.3449	0.445
X_{30}	0.5402	0.5709	0.4877	*0.8667*	0.5373	0.5724
X_{31}	0.5052	0.6239	0.4604	*0.8245*	0.4524	0.4575
X_{24}	0.3369	0.5026	0.5208	0.4546	*0.8566*	0.4447
X_{25}	0.5344	0.6151	0.433	0.5198	*0.8932*	0.51
X_{26}	0.54	0.5404	0.3967	0.4168	*0.8198*	0.4826
X_{15}	0.4434	0.6246	0.4057	0.489	0.4033	*0.7793*
X_{20}	0.4671	0.6695	0.596	0.6353	0.4983	*0.8854*
X_{21}	0.371	0.5178	0.4649	0.3956	0.4166	*0.7554*
X_{23}	0.5426	0.6788	0.4778	0.5719	0.5286	*0.8546*
X_{35}	0.5452	0.5902	0.4788	0.6004	0.4504	*0.8469*

Italics are indicators' loading with their associated latent constructs

8.4.3 Evaluation of the Structural Model

Given the measurement model was tested in Sect. 8.4.2 and the results generally support the measurement model, the measurement model can be used in the evaluation of the structural model's quality. As illustrated in Fig. 8.2, the structural model of this study consists of eight relationships between constructs. These relationships are represented by path coefficients or parameter estimates (*P*). The goal of the evaluation of the structural model is to assess the stability of these parameter estimates. Several criteria can be applied.

The evaluation is started with the coefficient of determination R^2. R^2 reflects the regression function's goodness of fit by measuring the share of the latent construct's explained variance. The higher R^2 is, the larger the share of the latent construct's variance explained. However, there is no common ground of the acceptable threshold of R^2. The judgment of acceptable values of R^2 depends on the specific research (Götz et al. 2010). In this study, R^2 for IU, which is the final dependent construct in the model, is 0.44. R^2 for other outcome constructs are 0.53 (FC); 0.46 (LA); 0.42 (BC); 0.56 (TA). Overall, these R^2 values of the model can be described as moderate (Chin 1998; Ringle et al. 2010; Hair et al. 2011).

The evaluation of the structural model is also based on the assessment of the significance of path coefficients (Chin 1998). The estimated path coefficients of the structural model of this study are provided in Table 8.18. These path coefficients can be regarded as standardized beta coefficients resulting from the least-squares method (Götz et al. 2010). The PLS-SEM approach uses bootstrap resampling procedure to test hypotheses about beta coefficients. The significance of each path coefficient is assessed by comparing the *t*-values from bootstrap resampling and critical *t*-values for a one-tailed test. As shown in Table 8.18, all but one structural path estimates are significant (at 0.1 level of significance) and in the expected direction. The estimate between FC and IU has *t*-value below 1.65, the critical *t*-value at 0.1 level of significance. Therefore, the hypothesized relationship is not supported. Overall, the theoretical model is supported with seven out of eight estimates being consistent with the hypotheses.

Table 8.18 Estimated path coefficients and *t*-value

Hypotheses		Parameter estimates		Standard deviation	*t*-value	*t*-critical value at 0.1 level
H_1	IC +→ FC	$P_{FC,IC}$	0.7311	0.0423	17.2785	1.65
H_2	IC +→ TA	$P_{TA,IC}$	0.7494	0.0380	19.7156	1.65
H_3	IC +→ LA	$P_{LA,IC}$	0.6773	0.0556	12.173	1.65
H_4	IC +→ BC	$P_{BC,IC}$	0.6457	0.0527	12.2468	1.65
H_5	FC +→ IU	$P_{IU,FC}$	0.1216	0.0843	1.4423	1.65
H_6	TA +→ IU	$P_{IU,TA}$	0.2803	0.1040	2.6958	1.65
H_7	LA +→ IU	$P_{IU,LA}$	0.2003	0.1008	1.9865	1.65
H_8	BC +→ IU	$P_{IU,BC}$	0.1940	0.1016	1.9087	1.65

In addition to R^2 measures and the significance of path coefficients, the predictive relevance of the structural model is used to assess the model's capability to predict each endogenous latent construct's indicators. The assessment is carried out by using a blindfolding procedure in which every *d*th data point part is left, out and the resulting estimates are used to predict the omitted part. The Stone–Geisser test criterion Q^2 is computed as follows:

$$Q_j^2 = 1 - \frac{\sum_k E_{jk}}{\sum_k O_{jk}}$$

where

E_{jk} the squares of the prediction errors, which represent the difference between true values of the data omitted from the blindfolding procedure and the predicted values;

O_{jk} the squares of the trivial prediction error provided by the mean of the remaining data from the blindfolding procedure;

J the observed endogenous measurement model; and

k the index for all indicators of the measurement model (Götz et al. 2010)

Q^2 can be provided in two forms: the cross-validated redundancy and cross-validated communality. In studies using the PLS-SEM approach, Q^2 as endogenous constructs' cross-validated redundancy measure value is more appropriate. In this study, the blindfolding procedure is applied for five endogenous constructs IU, FC, LA, BC, and TA. As shown in Table 8.19, all endogenous constructs' cross-validated redundancy is larger than 0, indicating that the model has predictive relevance (Fornell and Cha 1994; Chin 1998).

8.4.4 Interpreting Relationships

With estimated path coefficients, the relationships among target constructs are illustrated in Fig. 8.3. The sizes of these coefficients indicate that lack of transparency and accountability in infrastructure development has the biggest impact (0.28) on the inefficient use of public investments in infrastructure, whereas lack of

Table 8.19 Endogenous constructs' cross-validated redundancy

Total	SSO	SSE	1-SSE/SSO
FC	504	340.9448	0.3235
IU	378	275.2743	0.2718
LA	630	446.1687	0.2918
BC	378	265.2498	0.2983
TA	630	390.5696	0.3800

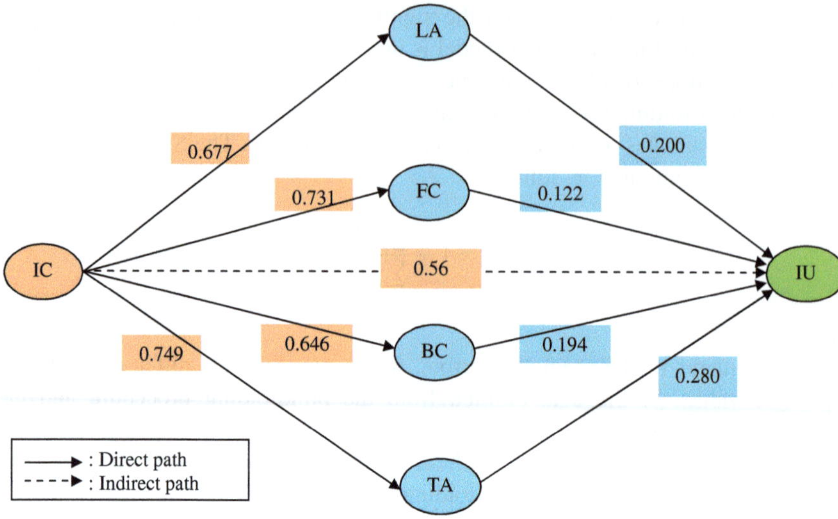

Fig. 8.3 Estimated structural equation model of factors affecting the use of public investments in infrastructure

an effective land acquisition framework is somewhat less (0.20) and insufficient building capacity of local firms has the smallest impact (0.194). Although the estimated parameter representing FC → IU is 0.122, the statistical test results show that there is a high probability that the actual is equal to zero. Inadequate forecasting capacity thus does not have significant impact on the inefficient use of public investments in infrastructure.

In addition, although there is no direct relationship between insufficient institutional capacity of the government, and the inefficient use of public investments in infrastructure, the indirect paths relate to insufficient institutional capacity of the government are considerable. Insufficient institutional capacity of the government has substantial impacts on lack of transparency and accountability in infrastructure development (0.749), on lack of an effective land acquisition framework (0.677), on insufficient building capacity of local firms (0.646), and on inadequate forecasting capacity (0.731). As a result, total estimated effect of insufficient institutional capacity of the government on the inefficient use of public investments in infrastructure is computed as a sum of these indirect paths:

$$\left.\begin{aligned} IC \rightarrow LA \rightarrow IU &= 0.677 \times 0.200 = 0.1357 \\ IC \rightarrow FC \rightarrow IU &= 0.731 \times 0.122 = 0.0889 \\ IC \rightarrow BC \rightarrow IU &= 0.646 \times 0.194 = 0.1253 \\ IC \rightarrow TA \rightarrow IU &= 0.749 \times 0.280 = 0.2101 \end{aligned}\right\} 0.56$$

The total estimated impact is substantial (0.56) and provide the evidence of that relationship as well.

8.5 Interviews and Discussion

8.5.1 *Profile of Interviewees*

After the questionnaire survey was conducted with some results, in-depth interviews were conducted. The interviews were conducted in Vietnam in July 2012. Fifteen professionals, who are holding key positions in the organizations involved in planning and executing national infrastructure development plans, were contacted (see Sect. 7.6 for sampling method). However, only seven of them agreed to be interviewed. The sample size satisfies the required minimum size of in-depth interviews for this study (see Sect. 7.6). Two samples of interview records are attached in Appendix 3. The profiles of these interviewees can be found in Table 8.20.

Based on the survey results, interviewees were mainly asked to explain why these factors occurred, and how to address them to improve the investment efficiency of funded infrastructure development in Vietnam. Using the thematic content

Table 8.20 Profiles of interviewees

Interviewee code	Designation	Organization's main business
G1	Director of Construction Department of Hai Duong province	Assisting People's Committee of Hai Duong province with construction management and planning
G2	Deputy Head of the Department of Infrastructure and Cities under the Ministry of Planning and Investment	Studying and synthesizing development strategies and master plans for infrastructure and cities; coordinating with other relevant units within the MPI to develop national and regional socio-economic development strategies
G3	Former Head of Finance and Planning Department under the Ministry of Construction	Compiling strategies, master plans, long-term plans, five-year plans, and annual plans for the development of the construction industry, as well as development plans for the Ministry's subordinate bodies
G4	Former Head of Government Authority for Construction Quality Inspection under the Ministry of Construction	Quality management for and inspections of construction works
P1	General Director of a private developer	Real estate development, construction of building, and civil engineering works
P2	Deputy Director of Vietnam Institute for Building Science and Technology	Building research and verification, and consultancy
P3	General Director of a private consultant	Consultancy (architecture, civil engineering, project management, and so on)

analysis, transcripts of interviews were first read and used to identify and check topics and aspects of contents. In the next stage, all contents of each topic were generally grouped together for direct reference for the discussion. Contents, including cases and examples, would be extracted from transcripts to link with the survey findings. Related documents were collected to illustrate various issues as well. The interviews are thus combined with the survey findings and discussed in the following sections.

8.5.2 Inadequate Forecasting Capacity

From the results of factor analysis, this factor consists of inadequate data, inherent problems in predicting the future and monetizing external and indirect effects, and lack of experienced forecasters. These issues are identified as technical issues that could affect forecasting accuracy (Flyvbjerg 2007). From the literature of development planning, forecasting has long been considered an important area in development planning as it guides decisions made by policy-makers (Lewis 1966). Inadequate forecasting capacity was thus hypothesized to have effects on the efficient use of public investments in infrastructure. However, the PLS-SEM analysis presented above leads to the rejection of such hypothesis. Several methods used to assess the stability of the structural model (see Sect. 8.4.3) suggested that the parameter coefficient of 0.2 between inadequate forecasting capacity and inefficient use of public investments in infrastructure is not statistically different from zero. It implies that data, techniques, and experiences in forecasting are not important variables in explaining the inefficient use of public investments in infrastructure in Vietnam. This is consistent with and complementary to indications in the growing literature on decision-making with respect to large infrastructure construction projects (Flyvbjerg 2007; Van Wee 2007).

The long tradition of planning has brought many benefits to the forecasting capacity of Vietnam. It allows the accumulation of data, experience, and knowledge, and the gradual improvement of forecasting techniques and methods. At the early stage of the planned economic regime, there was an understanding of the need for data collection. The Statistics Service, now the General Statistics Office under MPI, was established in 1946. Its major task is to collect and accumulate data for macroeconomic forecasting and planning. The planning system has also long developed a mechanism for forecasting. Macroeconomic forecasting, including forecasting for infrastructure development, lies with several institutions under the Ministry of Planning and Investment. For instance, forecasts by the Development Strategy Institute are used by the government to assist in the development of national socio-economic development master plans and strategies, including infrastructure development for periods from 5 to 20 years. Besides, macroeconomic forecasts by the Central Institute for Economic Management and the National Center for Socio-economic Information and Forecast are used by the government to assist in the development of areas such as economic policies and economic management strategies. At sectoral and local levels, forecasting is carried out by research and forecasting institutions under the sectoral ministries and local authorities. The establishment of such research and forecasting institutions suggests that the government of Vietnam has recognized the importance of forecasting and has invested resources to improve the government's forecasting capacity. Nevertheless, there still exists a substantial gap between forecasts and the actual results of infrastructure development planning (Nguyen and Dapice 2009). This can be explained that the forecasting capacity is available but its effective use might not be realized. G3 had experienced the difficulties when engaging in even sectoral forecasting, let alone national planning because input data submitted by enterprises were often manipulated. G2 shared that decision-making and planning in Vietnam was based on political leaders' will, rather than on forecasting or estimating. Political leaders could direct planners to prepare the master plans based on their subjective assessment. The interference of political leaders in planning (discussed further in the following sections) has rendered forecasting and estimating ineffective. Besides, P2 and P3 added that unanticipated economic disturbance (both external and internal) and natural disasters also affect the quality of forecasting. For example, the occurrence of the 2008 global financial crisis, the oil price increases over the last decade, and the Japanese earth quakes and tsunamis in 2011, all have effects on public investments in infrastructure in Vietnam. However, Vietnam has no control over these events, thus having difficulties in predicting their occurrence.

Inadequate forecasting capacity is thus not an important constraint to the current use of public investments in infrastructure in Vietnam. Improving forecasting capacity per se is not an important target. How to use these resources to provide good planning decisions seems to be a more important matter. As such the efficient use of public investments in infrastructure is not affected by the inadequate forecasting capacity but the lack of effective use of forecasting resources.

8.5.3 Lack of Transparency and Accountability in Infrastructure Development

This study hypothesized that lack of transparency and accountability in infrastructure development would be positively related to the inefficient use of public investments in infrastructure. That is because lack of transparency and accountability poses significant risks of corruption, and cost and time overrun, which in turn result in bad investments with poor quality of infrastructure and low economic returns (Kenny 2007; Sohail and Cavill 2008). Results of the PLS-SEM analysis and evaluation (see Sect. 8.4.3) have confirmed that lack of transparency and accountability is a statistically significant cause. The estimated parameter between TA and IU of 0.28 implies that lack of transparency and accountability in infrastructure development is largely responsible for the inefficient use of public investments in infrastructure in Vietnam.

From factor analysis, lack of transparency and accountability in infrastructure development in Vietnam is first reflected in the lack of the transparency and competitiveness in bidding and award. Transparency and competitiveness in bidding and award are very important as they ensure that a qualified firm is selected to fulfil project objectives. Bidding and award of public procurements are regulated by the Ministry of Construction, at the central government level and local construction departments, at local government levels. Since the industry's reforms in early 1990s, the regulatory framework of bidding and award of public works has been improved considerably. A number of related laws and regulations have been issued by the Ministry of Construction over the years. The most important milestones are the introduction of Construction Law in 2003 and Tendering Law in 2005. Bidding and award of all types of construction projects including funded infrastructure projects must be carried out in accordance with these laws and regulations (Ministry of Construction 2008). However, in practice open competitive tendering is merely a formality. The bidding procedures are primarily based on subjective assessments. G4 observed that qualifications of tendering firms were not sufficiently considered. With firms that meet minimum qualifications required, the only selecting criterion is tender price. With the acceptance of lowest tender price, there are no criteria to capture the difference in qualifications of tendering firms necessary to ensure firms' capability to fulfil the contract. Moreover, there are no guidelines to verify firms' tendering applications and to prevent fraud. As a result, firms are primarily selected based on the lowest evaluated tender price. Contract winners are often those having strong relationships with relevant government ministries or procurement officials. Furthermore, the Tendering Law does not forbid tender price to be lower than a floor tender price. G3 and G4 indicated that many Chinese firms took advantage of the regulation and produced strategically under-evaluated tender prices to get the jobs. However, without adequate resources and the required capability, many Chinese firms could not complete these projects. G1 and P3 also shared that firms could involve collusive agreements with others to manipulate bidding prices in the bidding processes. According to a recent report on building quality management in transport infrastructure projects issued by the Transport Engineering Construction and Quality Management Bureau of Vietnam (TEC and QMB), these practices pose a threat to the quality of construction projects. After the contract has been awarded, firms would lower the quality standards to lower contract costs (Transport Engineering Construction and Quality Management Bureau 2011).

In addition, under the Tendering Law, selecting contractors, suppliers, and consultants for infrastructure projects with the contract valuation below a certain level can be carried out by limited tendering or direct appointments of contractors (National Assembly of Vietnam 2005). This regulation creates improper incentive mechanism for avoiding the open tendering process by adjusting the valuation of contracts or dividing procurement into smaller contracts (Transport Engineering Construction and Quality Management Bureau 2011).

Lack of transparency and accountability in infrastructure development in Vietnam is also reflected in the complicated and bureaucratic administrative system that exists throughout the planning and execution stages of infrastructure projects. The development usually involves conceptual proposal, pre-feasibility study, feasibility study, detailed engineering design, appraisal, and final approval. During the process, nearly every decision, which was within the capacity of Project Management Units, required some form of review and approval by a number of relevant ministries and local authorities before finally being approved by the Prime Minister. Project design alone has to go through a number of steps of verification and approval by related government agencies (Ministry of Construction 2013). Project costs are calculated using state norms and quota systems for labour and materials and so on. Deviation from the norms is required to be approved by relevant government offices at intermediate levels before moving to the Prime Minister's Office (Prime Minister's Office of Vietnam 2009b). All interviewees reported that bureaucratic procedures have caused significant delays and raised costs in the implementation of public works. However, the government is rarely held accountable for its decisions and actions when infrastructure projects fail to achieve project objectives. As described by G1:

> Many meetings would be held before a project proposal was approved. Attending these meetings were a number of representatives from various government agencies. But, no single government official had to take the responsibility for the decisions made. For instance, a number of committees would be set up to study project proposals. When a project was completed, a committee, whose chairman would have never been to the project site, was also responsible for the issuance of the official acceptance of work. That is because the issuance is merely a formality. Whether the quality of work is acceptable is not responsibility of the chairman.

Besides a bureaucratic administrative system, collective responsibility is still a prevailing cultural issue in Vietnam. Vietnamese people would expect other group members to protect them from any individual responsibility, and vice versa (Low and Dang 2008). Low and Dang (2008) also described collectivism as one of the major unfavourable conditions for organizations to effectively deal with each other in Vietnam.

P1 explained that the combination of many roles of government as decision makers, regulators, supervisors, clients, and so on prevents the detection of fraud or corruption around the actions of government officials. For example, quality inspection for all funded infrastructure projects is carried out by several government agencies such as the State Authority for Construction Quality Inspection (under the Ministry of Construction) and Transport Engineering Construction and Quality Management Bureau (under the Ministry of Transport). Using the complicated rules and procedures, government officials can involve collusive agreements with consultants and contractors to cover poor-quality construction practices and outcomes. This is compounded by the poor enforcement of the existing legislation and regulations. G3 indicated that the existing legislation and regulations were not equipped with effective tools to protect whistle-blowers and punishments stringent enough to prevent corruption. According to a report issued by the Government Inspectorate of Vietnam, discovering fraudulent and law-breaking practices in the construction sector are difficult due to the collusive agreements and complicated association among the parties involved. In addition, when discovered, those responsible for the law-breaking practices are not promptly punished due to bureaucracy in administrative procedures. These issues have significantly dampened anti-corruption efforts as well as encouraged harmful practices in the sector (Government Inspectorate of Vietnam 2005). The lack of transparency and accountability is thus an important constraint on the efficiency of funded infrastructure development in Vietnam.

8.5.4 Lack of an Effective Land Acquisition Framework

Without an effective land acquisition framework, land acquisition for infrastructure development can be costly and time-consuming (Priemus 2010). Thus, this study hypothesized that lack of an effective land acquisition framework would positively affect the investment inefficiency of funded infrastructure development. Based on the PLS-SEM analysis and evaluation, the underlying cause is statistically significant (see Sect. 8.4.3). The relationship between LA and IU is illustrated by the estimated parameter of 0.2. The results suggest that an improvement in the land acquisition framework can mitigate the risk of investment inefficiency of funded infrastructure development in Vietnam.

Land acquisition in Vietnam is carried out in accordance with the provisions of Land Law. According to the first Land Law issued in 1987, there was no private landownership in Vietnam. The government representing the people owned and distributed land to users. The government had the right to acquire land that had been distributed and land users had no right to transfer land. However, since 1988 more rights over the land have been given to land users. The Land Law issued in 1993 officially gave land users the rights of land transfer and rental, which made land an asset to users (Han and Vu 2008). Since then, the issue of compensation has become the core element of land acquisition. According to the latest Land Law issued in 2003, land users have the rights to claim for compensation when land is acquired. In order to have the rights and to avoid any disputes, land users are required to have legal titles to the land. However, up to 1999, urban households without land-use certificates still accounted for about 90 % of total urban household (Quang and Kammeier 2002). Bureaucracy in settling land claims and disputes is regarded as a major contributing factor to inefficiency. Various legal documents issued by different government offices are required before a land-use certificate would be issued. High tax and legal scrutiny required for transfer of land-use titles also prevent land users from applying for land-use certificates (Kim 2004). Besides, overlapping and inconsistency between land-use regulations make land claims and disputes settlement more complicated (Thanh 2012).

There are two types of procedures for compensation. For large public projects, the government would set up a site-clearance committee to acquire the land. For private projects, developers would have to negotiate to buy land from the residents. Valuation for compensation is based on the state price framework. Prices listed in the price framework are required to be close to market prices and updated annually.

In addition, the prices can be adjusted within a limited range depending on the conditions of the projects (National Assembly of Vietnam 2003). In principle, based on the price frame, government officials and developers would determine the amount of compensation. In practice, the process is more complicated. In private projects, private developers could initially value the land on the basis of the government price framework. However, transaction prices were often higher. That is because land users have the right to decide the price at which they want to sell their land. Many land users abuse the right to force developers to buy their land at rates much higher than the rates set by the government. In addition, the mechanism encourages speculative purchases of land in the areas which were announced for acquisition. Purchasers of the land have high expectation that the price would increase dramatically when the infrastructure were developed. As a result, compensation for land users in these areas would have to be much higher than generally accepted (Han and Vu 2008).

On the contrary, in land acquisition for public projects, the government is both the takeover and the valuer. Government officials decide compensation rates based on the government price framework, which were regarded as comparable with market value. Land Law 2003 stipulates that local authorities have to build price frameworks closely to market prices under normal conditions, and the price frameworks would have to be adjusted when there is a large difference between the frameworks and market prices. The law also specifies that assistance measures for resettlement would be provided so that resettlement sites should offer conditions which are at least equal to the previous one (National Assembly of Vietnam 2003; Prime Minister's Office of Vietnam 2009a). Nevertheless, the extent to which price frameworks are considered to be close to the market price, or the extent to which resettlement sites is equal to the previous one are not defined in the law. Moreover, there is also no independent monitoring system for the implementation of resettlement to ensure that adequate assistance is provided. Having no right to negotiate with the government, land users have to accept the compensation packages, which are often lower than what they expected. In many cases, rural land users were poorly paid because they did not have any clear title to the land. With inadequate compensation packages, land users usually have to relocate far from the city or employment centres. For rural land users, without adequate assistance, they also faced difficulties in settling down to a new way of life or finding new jobs. This has created a much dissatisfaction among those whose land was being taken away (Kim 2011).

In both compensation procedures, there is a risk that a compensation package would become increasingly unacceptable to one of the parties. While in private projects, compensation rates would be driven to an unreasonably high level, and there is always a disagreement over the adequacy of compensation packages in public projects. The situation is worse in public projects since the government do not often have sufficient public financial resources to offer a higher compensation rate in order to clear a site more quickly. As a result, a compensation assessment would be difficult to be reached, which in turn slow down the project development, causing significant delays and cost overruns.

Interviewees regarded land acquisition as the major reason for time delay, especially in projects located in urban areas. In many cases, project construction could not start as planned because of the difficulties in the land acquisition process. G1, G4, and P2 observed that execution of many infrastructure projects had to be cut into parts. Project construction would start first in the area where the land acquisition was settled. Construction of the remaining parts of the road would have to wait until the compensation deals for the remaining area were reached. P2 mentioned the construction of the Cau Gie–Ninh Binh highway (a section of the North–South highway project) as an example. The development of the Cau Gie–-Ninh Binh highway was approved by the Prime Minister in 2004. The total length of the highway was planned to be 50 km. In 2011, after 7 years of construction, only about 20 km of the highway (the part connecting Cau Gie and Phu Ly) was opened for use. The construction of the remaining of the highway had faced a number of issues in land acquisition, which were finally settled in March 2012 (Tinh 2012).

G1 suggested that in order to accelerate the land acquisition process in Vietnam, land acquisition, and compensation for public purposes in Vietnam should be separated, in which land should be acquired first for project development, and compensation deals would be settled later. However, the measure would leave no room for land users to challenge the validity of land acquisition. Meanwhile, G4 suggested that the first step to solve land acquisition issues is that the government should have clear spatial plans for infrastructure development, based on which land would be set aside in advance. In the case of land acquisition for public projects involving resettlement, public support for public projects is often low. P2 and G4 recommended that the government should do more to educate the public about the citizens' obligations to serve national interests. Besides, compensation packages should be adequate to cover the loss of business and ensure a smooth transition to a new way of life for displaced land users. The interviewees emphasized that improving transparency and accountability is necessary to ensure that in practice, adequate compensation and assistance would be provided.

Concerning public finance resources for compensation, G4 emphasized that the key is to capture a proportion of the increase in land value associated with infrastructure improvements. Besides raising revenue for the government for compensation, capturing the increase in land value has a potentially larger effect to achieve equality between those who suffer and those who benefit from infrastructure improvements, thus gaining public support. In an effort to capture the increase in land value, G4 shared that the local government of the central city of Da Nang has taken an initiative. G4 explained that in Da Nang when a road, for instance, was planned to be built in Da Nang, not only the road site but also land in the immediate vicinity would be acquired. The local government would use the revenue from selling the surrounding areas for private land development to compensate the sitting tenants and to finance the road development project. However, the method might not be applied in Hanoi where houses of a number of government bureaucrats are located, said G4. There were also cases where local governments would exchange private land development approval for the public financing of public infrastructure including compensation costs. However, the method was criticized for its low level of transparency and competition, and thus heightening the potential risks for corruption (Kim 2011).

The issues in compensation procedure for public projects highlight the need for a negotiating mechanism to make the market price of land play a correct role in land acquisition compensation. The government, being both the takeover and the valuer, could not ensure a fair compensation valuation for land users. Controversies over compensation packages are the root cause of increasing claims and disputes in land acquisition. Meanwhile, the issues in the approach taken in private projects underline the need for law enforcement to regulate the price of land acquired from growing wildly, which could affect the whole land market. Besides, the lack of transparency and accountability in the valuation methods and compensation process and bureaucracy in land disputes and claim settlements are further evidence that the land acquisition framework for infrastructure development in Vietnam is in need of improvement.

8.5.5 Insufficient Building Capacity of Local Firms

The results of the PLS-SEM analysis show that insufficient building capacity of local firms positively impacts the inefficient use of public investments in infrastructure. The test of significance also confirms that the parameter of 0.194 between BC and IU is significantly different from zero (see Sect. 8.4.3). In other words, capacity of local firms is identified as a constraining factor of the efficiency. The local infrastructure construction sector is dominated by a few large state-owned construction firms. The rest of the sector is made up of several thousand small-to-medium construction firms (Kenny 2007). Involvement of foreign construction firms is greater in large-scale infrastructure projects (often funded by ODA). However, participation of these foreign firms is mainly through a joint venture with local firms. Local construction firms are often criticized for the poor level of efficiency and quality of work (Long et al. 2004; Low and Dang 2008). The interviewees reported that most of the parties involved are responsible for these problems.

First, developers of infrastructure projects under the responsibility of the central governments as specified in national infrastructure development plans are a few large state-owned corporations or the relevant ministries. These developers are represented by a Project Management Unit, a temporary organization set-up to be responsible for project development. G1 shared that in large-scale infrastructure projects which required greater technical expertise, assigned developers or PMUs often did not have the required capacity and level of professionalism. The issue is also reported by TEC and QMB (2011). Most PMUs employed supervision consultants representing them to manage projects. Being overly dependent on supervision consultants, government officials of PMUs were responsible for the lack of involvement in the project life and slow decision-making. P2 and P3 commented that developers (PMUs) were often responsible for interruptions in funding. Uncertainties over the government budget allocation, lengthy and time-consuming payment procedures created cash flow problems for contractors and consultants to implement projects on time. Another concern is PMUs' lack of profit motivation. PMUs are responsible for implementing infrastructure projects without any financial liability. G1 explained that after the delivery of physical facilities, PMUs would not be tied to any financial liability. Costs of construction, operations, and maintenance would be the responsibility of other relevant government agencies, which later run the infrastructure facilities. As a result, PMUs lack a commercial focus to cut the costs of construction, as well as of operations and maintenance by setting appropriate design and technical standards for infrastructure facilities.

Supervision consultants are employed to manage project execution. Supervision consultants represent clients (PMUs) to supervise project execution of contractors including certifying volume and quality of work done by contractors, approving construction methods for contractors, and proposing solutions for technical issues on site. In Vietnam, local supervision staff play an important role in quality management, assisting clients to ensure quality throughout projects. According to TEC and QMB (2011), even when foreign supervision consultants are employed (such as in ODA funded projects), foreign engineers accounted for a small portion of the workforce and mainly acted as chief engineers. The rest of the workforce was made up of local engineers. However, there are concerns with the capacity of local supervision consultants to adequately fulfil their roles in large funded infrastructure projects, including lack of experience in contract management, lack of quality control over contractors' performance and slow response to site-related issues. G1 commented that since clients and PMUs used public funds without any financial liability, supervision consultants would be poorly selected. In addition, G4 shared that:

> The collusion between supervision consultants and contractors to overlook poor quality materials and works is one of major issues affecting project quality. Although local supervision consultants were able to provide supervision staff in quantity, they lack highly qualified and ethical professionals in their workforce. It has been difficult to mobilize and retain skilled staff in the supervision consultancy profession. Low income is considered the major cause of the issue. Despite its important role in quality assurance, the supervision consultancy profession has not received adequate attention in the industry.

According to TEC and QMB (2011), the workforce is mainly provided by supervision centres of relevant universities, institutes, government agencies, and some private organizations. There are organizations providing training, certification, and registration for supervision consultants. However, there are no system to guarantee their consistency and standard procedures to ensure quality and ethics of supervision consultancy.

During the design stage, designers and engineers employed to design the projects are responsible for poor quality of feasibility study and design, and impractical engineering solutions and technologies (Long et al. 2004). The report issued by TEC and QMB (2011) indicated that except for a few large design and engineering consultants, a large number of small local design and engineering firms have been established in recent years to meet increasing market demand. However, without long-term development plans, these firms were lagging in the market and incapable of handling large projects. G1 reported that the local industry lacks experienced and highly qualified designers and engineers for undertaking feasibility studies at macro-level, preparing master plans, proposing engineering solutions and technologies that are within the capability of local contractors. G1 and P3 added that local designers and engineers are often overly influenced by local government authorities (clients), instead of defending their proposals. Besides, they tend to rely on standard engineering solutions, rather than putting in efforts to explore alternative designs. As a result, the selected design and engineering solutions would not be cost-effective and optimal to specific conditions and complexities of projects. P3 quoted a case where a foreign engineering solution, stone-mastic asphalt (SMA), was selected to upgrade the surface of the Thang Long Bridge (connecting Hanoi and Noi Bai International Airport). The technology was first developed in Germany in the 1960s. This was the first time the technology was used in Hanoi. However, a few months after the project was completed (at the end of 2009), there were a number of serious cracks on the bridge's surface. According to a report by the State Authority for Construction Quality Inspection (Thinh 2011), it was concluded that the incompatibility between the newly stone-mastic-asphalted surface and the existing steel structure of the bridge was the main cause of the incident.

Quality management was poorly implemented by local design and engineering consultants. G1 observed that they lack in-house resources and capacity for design coordination, review, and checks. Quality of design could not also benefit from design verification. TEC and QMB (2011) reported that design verification mainly focused on criticizing design weaknesses rather than on providing suggestions to optimize design. There is no local firm specializing in verification. Design verification was mostly done by relevant institutes or universities. Besides, P3 commented that the poor quality of input data have significantly affected the quality of design. For example, inadequate data from topographical surveys could result in design variations. Topographical surveys are the basis for engineering design. However, G4 shared that project funding for topographical surveys in Vietnam was limited. Topographical surveys were often not sufficiently carried out and checked in order to provide accurate and detailed results. Errors in the survey results were only discovered during the execution stage. Consequently, design with poor quality had to be revised or added during project execution, resulting in project delay and waste as well as affecting the overall project quality (Government Inspectorate of Vietnam 2005; Transport Engineering Construction and Quality Management Bureau 2011). P2 commented that unanticipated weak soil foundation was the main cause for the collapse and settlement of roads and bridges during construction in Vietnam. P2 quoted the case of the Can Tho Bridge (connecting Can Tho City and Vinh Long province in the south of Vietnam) where the collapse of two side spans of the bridge during construction in 2007 was due to the unanticipated weak soil foundation.

During the construction phase, local contractors were found to be responsible for financial difficulties, inaccurate project planning and scheduling, poor construction quality, inadequate, and ineffective coordination among parties, poor site management and inadequate use of modern equipment (Long et al. 2004). Besides cash flow problems caused by clients, financial difficulties of local contractors were attributable to their own internal weaknesses. As reported by TEC and QMB (2011), at the management level, local contractors lacked the expertise in corporate finance management, including the right skills to carry out proper calculations of funds for operational and developmental purposes, and project cash flow analysis. Project financial resources were inappropriately used or diverted for other projects, resulting in serious cash flow problems during project execution. Cash flow and work flow are poorly managed, resulting in inefficient use of inputs. Relating to the development of human resources, most contractors do not have any human resource management policy. As explained by G1, they hired an informal labour force and only provide poorly on-the-job training to save on costs of production. TEC and QMB (2011) also reported that at the technical level, local firms lacked professional and skilled staff capable of managing large infrastructure projects. Many contractor staffs are not familiar with the latest project management techniques and emerging technologies. Local contractors also encountered the limitation of equipment resources in large infrastructure projects. Investments in modern technology and equipment are limited due to financial constraints (Transport Engineering Construction and Quality Management Bureau 2011).

Concerning building quality, local firms lacked a quality management structure and culture. Although many firms have obtained ISO 9001:2000 certification, they only implemented the ISO system in their offices rather than on the construction site. Policies and procedures for health and safety in construction work were not taken seriously (Transport Engineering Construction and Quality Management Bureau 2011). G1 shared that most contractors depend on other parties including designers, supervision consultants, and related government agencies for quality management. Investments in quality control and quality assurance in these firms were not prioritized. Subcontracting practices in funded infrastructure projects also make the issue worse. Many public projects awarded to qualified contractors were subcontracted several times over. However, contractors often failed to provide adequate supervision of lower-level contractors, leading to substandard work. Another problem with project quality is related to fraudulent practices. G1 and P3 observed that:

> Very few firms understand that time is money, and focus on improving construction practices to save time and money, thus reducing costs and raising revenues. Instead, most firms increase profit margins and reduce construction costs through poor quality construction practices and material theft. Moreover, firms that had to put in low bid prices or involved some forms of corruption to get the jobs often wanted to compensate for the low prices or the corruption expenses through lower quality standards.

Vietnamese construction firms lack the expertise and resources in management, finance, and technology to guarantee quality and professional competence of large infrastructure projects. The capacity constraint of local construction firms is evident from poor-quality, poor construction planning and project management, and ineffective utilization of input resources. Local construction firms suffer from a serious lack of professional management structure and culture. Quality management system is generally inadequate. These supply constraints faced by the local infrastructure construction sector are at the project level and are subject to the capacity of individual firms. However, structural equation modelling (SEM) in this study revealed that there was a certain relationship between building capacity of local firms and the efficiency of funded infrastructure development of Vietnam. Capacity constraints of local firms affected their own efficiency, which eventually resulted in the deterioration of efficiency of funded infrastructure development. The relationship suggested that the efficiency of funded infrastructure development can be increased through increasing capacity of local construction firms. Increasing building capacity of local construction firms can benefit both the government and local construction firms. Thus, there is a need for the government to consider the area for improving the building capacity of local firms when formulating the national infrastructure development policy. Constraints that local firms encountered call for actions to be taken in tackling human resources and skill development issues, investing in quality management, modern technology and equipment, building professional working culture, and offering more financial means for local firms to sustain efforts on large civil works.

8.5.6 Insufficient Institutional Capacity of the Government

Insufficient institutional capacity of the government was hypothesized as the root construct affecting other factors. The PLS-SEM analysis and the tests of significance indicated that insufficient institutional capacity of the government had direct effects on other factors including inadequate forecasting capacity, lack of transparency and accountability, lack of an effective land acquisition framework, and insufficient building capacity of local firms (see Sect. 8.4.3). This creates an indirect relationship between insufficient institutional capacity of the government and the inefficient use of public investments in infrastructure. The combined effect of IC on IU is 0.56, which placed insufficient institutional capacity of the government in the critical position among factors affecting the use of public investments in infrastructure in Vietnam. This is consistent with the literatures on development policy-making and the role of the government (Todaro and Smith 2009).

The government of Vietnam has been structured relatively vertical. One of the weaknesses of the institutional structure is the lack of horizontal coordination. The causal explanation can be identified in the literatures on governance and economic development. The increasing complementaries between economic activities require a high degree of coordination across several policy spheres. This requires capabilities that emphasize inter-relationships, interdependencies, and a holistic rather than a hierarchical approach. The need to manage these interdependencies across sectors poses difficulties for the traditional form of institutional structure. Traditionally, the government is organized with many specialized and non-overlapping departments resulting in fragmented planning and policy-making, and uncoordinated actions. Lack of policy cooperation and coordination can thus pose high risks of inefficiency (Christensen and Lægreid 2007). At the national planning level, the lack of horizontal coordination in infrastructure development is reflected in the lack of inter-ministerial and intersectoral coordination. Given the interdependencies and linkages between infrastructure and other sectors as well as the economy as a whole (see Chap. 3), infrastructure planning for trade and economic development involves various sectors such as trade, land, construction, and infrastructure service technology. National development planning in Vietnam as described in Chap. 6 is a top-down and bottom-up process led by the Prime Minister and coordinated by the MPI. Top-down planning is formulated by a strategic planning subcommittee, which is staffed by top civil servants from Party institutions, government agencies, and national research institutes and a drafting committee led by the MPI. Bottom-up planning involves the policy needs and interests of line ministries and local governments. The MPI is tasked to drive planning and overall monitoring and evaluation from the centre of the government (Pincus and Nguyen 2004). However, there has not been enough systemic effort to ensure that the visions and strategies of the sectors and local governments mesh with one another. For example, as G3 and G4 explained, although every line ministry or local government had its own master plan for building airports, and seaports in the local area, there was no organizer to compile the plans. There is no central system of horizontal coordination between ministries and between lower levels of government, which make it difficult for any individual government department to collaborate across institutions. In the absence of such a system, the MPI's coordinating function is largely limited to monitoring separate, sometimes contradictory, policy agendas adopted by individual sectors and regions (Government Inspectorate of Vietnam 2005). G4 quoted a case of the Can Tho International Airport, in which the airport has been built without a plan for attracting international flights to the airport. Major seaports were complete without suitable passageways for accommodating ships, said G4. G1 also described the case of the Thanh Tri Bridge:

> Because the design of Thanh Tri Bridge (the largest among 7 bridges across the Red River in Hanoi) was done within the transportation sector, it was quickly readied at the time of decision making. The construction of the bridge started in 2002. A lot of money thus had been invested in building it. However, the access roads to the bridge were not put on the table because of the difficulties in land acquisition, which involved sectors and the local government other than the transportation sector. As a result, after the bridge had been completed for more than 2 years, the access roads to the bridge were still under construction, and the bridge could not be opened for use.

G1 and G2 also indicated that there is a lack of connection between planning and allocation of resources. Allocation of investment budget is often inconsistent with the long-term plans. G1 revealed that:

> Although it is called Ministry of Planning and Investment, more than 80 % of the MPI's work involves allocating investment budget. The MPI's development planning function only accounts for a small part… Development plans require a long term vision, e.g. for 5 years, 10 years. But, what planners mostly do is to allocate the investment budget on a

> yearly basis… Investment budget to be available in the coming fiscal year would be allocated to short-term needs arising at the time of decision making, rather than being linked to the long-term goals specified in development plans. Long-term development plans are prepared just for the sake of planning. Many investments were planned and approved by various government levels even though there is no corresponding plan of allocating capital investments… As a result, investments were scattering and incomplete, and things that must have been done first were done later, and vice versa.

As explained by G4, the excessively spread and disjointed investments would be attributable to the lack of a long-range vision of the planners. The lack of competent and qualified planners has constrained the government's ability to set clear economic development objectives, formulate strategies, and prioritize action plans. As described in Chap. 6, development plans in Vietnam were prepared by the MPI and approved by the National Assembly. However, G1 reported that most members of the National Assembly had a limited knowledge of construction investments. Besides, as reviewed in Chap. 6, there was little evidence of consultations in the public sector in the planning process in Vietnam. G1 and G2 shared that public influence on the final decisions made by related government agencies were not the major concern of the planners. Instead, planning was based mainly on the will of the political leaders, commented G2. However, G1, G2, and P1 indicated that in lacking a commitment to national goals, political leaders and government bureaucrats in Vietnam focused on taking advantage of the state budget available at the time of decision-making, rather than supporting the national development goals. P3 also explained that political leaders and government bureaucrats could manipulate budgeting decisions so that the infrastructure development plans with overestimated benefits, underestimated costs, and unfeasible schedules could be approved. Besides, in pursuing their personal, sectoral, or regional interests, they lacked a long-term vision in development planning. The tendency of politicians and civil servants to pursue self-interests could therefore result in poor judgment, poor quality of planning, and bad policy choice, leading to resource wastage and inefficiency of infrastructure development in Vietnam. Moreover, during the planning process, ex-ante assessment of infrastructure development plans by independent-review bodies and the scientific and professional community were considerably poor. P1 and G2 indicated that decision-making and assessment were both implemented by the government. The interviewees thus suggested the setting up of an independent monitoring and evaluation mechanism, and reducing the interference of political leaders in planning were necessary throughout the process of planning and approval for infrastructure development.

During the implementation of infrastructure development plans, the institutional capacity of Vietnam also lacks a high-powered agency responsible for the success of the entire delivery of infrastructure assets. The functional and fiscal decentralization associated with public investments in infrastructure has been increasing in recent years. The actual implementation of national infrastructure development plans in Vietnam rests with a number of line ministries and lower levels of government. More specifically, the enforcement of construction and land-use regulations are under the authority of local governments (Han and Vu 2008). However, as revealed by G2 and G3, the declining central control and increasing local autonomy in infrastructure development before the managerial capacity of local governments become sound enough have resulted in wasteful investments, abuse of public funds at the lower level of government. G2 and G3 commented that a central capacity to drive priorities and ensure quality control is necessary to provide an effective mechanism for implementing national infrastructure plans. Cabinet committees (for example, Steering Committee on master planning for national development of electricity; and Steering Committee on key transportation projects) could be established to organize and coordinate the implementation of these national infrastructure projects (Government Web Portal of Vietnam 2012). In most instances, the interdepartmental teams of officials are, however, linked to one another only when updates have to be provided. Moreover, the performance monitoring and evaluation were poorly carried out. There are no designated measurable criteria to evaluate the real costs of the infrastructure delivery. Similarly, there is no established procedure for ex-post assessment to verify whether infrastructure projects meet objectives. Reports produced by government agencies (Government Inspectorate of Vietnam 2005; Transport Engineering Construction and Quality Management Bureau 2011) are often long on describing the current status of related sectors but short on analysis and practical policy implications. Accordingly, remediation or acts to ensure that improved financial and physical performances actually take place in the delivery of national infrastructure assets was often vaguely suggested in these reports.

The last issue relating to institutional capacity of the government is the changing and inconsistent laws and regulations in infrastructure construction. Governments directly influence the public construction sector by setting rules for development that govern contractual relationships. The common features of legal systems in developing countries are that rules are constantly changed and inconsistent. The uncertainty caused by the inconsistency in policies, changes in laws and regulations could pose a high degree of risk to the industry's development and growth. Difficulties in complying with changing and inconsistent laws and regulations are regarded as the main obstacles in improving the industry's performance (Ofori 2000). Following rapid economic development since the country liberalized its market in 1986, the Vietnamese infrastructure construction sector has also increasingly expanded. Over the years, innumerable laws and regulations for governing the sector have been enacted (Ministry of Construction 2008). However, there is still a need for a unified legal framework. In an interview with the Vietnam Economic Times (Anh 2011), the vice chairman of VFCEA, Dr. Pham Sy Lien, shared that there are a number of inconsistency, duplication, and even contradiction between the Construction Law introduced in 2003 and others introduced more recently. The chairman explained that since there was an urgent need for a governing legal framework, the Construction Law was created within 6 months and set-up to cover all types of construction activities and participants. At the same time, the rapid pace of legal development left many issues unanswered. Similarly, the Tendering Law rigidly applies international formats without taking into consideration the local conditions, thus overlooking many important issues to protect the interests of the local participants. As a result, amendments and new laws and regulations are constantly enacted to deal with these issues. Although they have their respective emphasis, many of them are inconsistent with the existing statutes and codes (Anh 2011). G2 asserted that laws and regulations on construction investments and financing in Vietnam were frequently changed and inconsistent, causing fluctuations in prices and inflation rates in Vietnam, which in turn raised land compensation rates, material prices, salaries, and consequently capital investments. Eventually, these problems could affect the stability and continuity of the implementation of the country's development plans in general and the infrastructure sector in particular.

By setting national policy agenda, making national policy decisions, guiding law-making process, as well as monitoring and guiding policy implementation, an adequate institutional capacity of the government that generates good planning and sensible policies get to be implemented. Without an adequate institutional capacity, the government would move policy agenda unrealistically or set planning policies that do not reflect the interests of a majority of the population. Besides, institutional capacity specifies an operating environment and a set of rules for the implementation of government policy. It provides a framework to ensure the good performance of various actors in carrying out central policies. The empirical results indicate that an inadequate institutional capacity of the government influences forecasting, transparency, and accountability, as well as land acquisition and capacity of local firms, which in turn directly influence the efficient use of public investments in infrastructure. This suggests that inadequate institutional capacity of the government is at the root of investment inefficiency of funded infrastructure development in Vietnam. An adequate institutional capacity of the government is thus a precondition for improving the efficient use of public investments in infrastructure. This is consistent with the literature on economic developmental pointing to development planning and the role of the state.

8.6 Summary

Analysing the data collected, the chapter shows that efficiency of public investments in infrastructure in Vietnam is falling. Factor analysis indicates that factors critical to the current use of public investments in infrastructure in Vietnam are institutional capacity of the government, transparency and accountability, land acquisition, and building capacity of local firms. Among these factors, institutional capacity of the government is at the root of the efficient use of public investments in infrastructure. The results presented in this chapter will be used to develop the research framework in the next chapter. Discussions of the related problems and challenges in this chapter will also be extended to provide policy implications and recommendations in the following chapter.

References

Afonso, A., Schuknecht, L., & Tanzi, V. (2005). Public sector efficiency: An international comparison. *Public Choice, 123*(3–4), 321–347.

Anh, B. (2011, October 3). Tendering Law does not protect local firms. *Vietnam Economic Times*. Retrieved from http://vneconomy.vn/20110930081227725P0C9920/luat-dau-thau-khong-bao-ve-nha-thau-trong-nuoc.htm.

Arkadie, B. V., & Mallon, R. (2003). *Vietnam: A transition tiger?* Australia: Asia Pacific Press.

Chiang, Y.-H., Anson, M., & Raftery, J. (2004). *The construction sector in Asian economies*. New York: Spon Press.

Chin, W. W. (1998). The partial least squares approach for structural equation modeling. In G. A. Marcoulides (Ed.), *Modern Methods for Business Research* (pp. 295–336). Mahwah: Lawrence Erlbaum.

Christensen, T., & Lægreid, P. (2007). The whole-of-government approach to public sector reform. *Public Administration Review, 67*(6), 1059–1066.

Flyvbjerg, B. (2007). Megaproject policy and planning: Problems causes, cures. *Environment and Planning B: Planning and Design, 34*, 578–597.

Fornell, C., & Cha, J. (1994). Partial least squares. In R. P. Bagozzi (Ed.), *Advanced methods of marketing research* (pp. 52–78). Cambridge: Blackwell.

Götz, O., Liehr-Gobbers, K., & Krafft, M. (2010). *Evaluation of structural equation models using the partial least squares (PLS) approach handbook of partial least squares* (pp. 691–711). Berlin: Springer.

Government Inspectorate of Vietnam. (2005). *Waste in funded infrastructure investment: Who is responsible for It? (in Vietnamese).*

Government Web Portal of Vietnam. (2012). List of steering committees. Retrieved Jan 20, 2013, from http://www.chinhphu.vn/portal/page/portal/chinhphu/cactochucphoihopliennganh?governmentId=2856&organizationTypeId=1.

Hair, J. F., Black, W. C., & Babin, B. J. (2010). *Multivariate data analysis: A global perspective* (7th ed.). London: Pearson Education.

Hair, J. F., Ringle, C. M., & Sarstedt, M. (2011). PLS-SEM: Indeed a silver bullet. *The Journal of Marketing Theory and Practice, 19*(2), 139–152.

Han, S. S., & Vu, K. T. (2008). Land acquisition in transitional Hanoi, Vietnam. *Urban Studies, 45* (5–6), 1097–1117.

Kenny, C. (2007). *Construction, corruption, and developing countries* (Policy Research Working Paper No. 4271). Washington, D.C.: World Bank.

Kim, A. M. (2004). A market without the right property rights. *Economics of Transition, 12*(2), 275–305.

Kim, A. M. (2011). Talking back: The role of narrative in Vietnam's recent land compensation changes. *Urban Studies, 48*(3), 493–508.

Lewis, W. A. (1966). *Development planning: The essentials of economic policy*. London: Allen & Unwin.

Long, N., Ogunlana, S., Quang, T., & Lam, K. (2004). Large construction projects in developing countries: A case study from Vietnam. *International Journal of Project Management, 22*(7), 553–561.

Low, S. P., & Dang, T. H. G. (2008). *Project partnering in Vietnam's construction industry. The Singapore engineer* (pp. 16–22). Singapore: Institution of Engineers.

Ministry of Construction. (2008). *The history of construction industry of Vietnam*. Hanoi: Construction Publishing House.

Ministry of Construction. (2013). *A new decree on issuing construction permission and approval: A step towards administrative reform (in Vietnamese).*

Mody, A. (1997). Infrastructure delivery through central guidance. In A. Mody (Ed.), *Infrastructure strategies in East Asia: The untold story* (pp. xi–xxvii). Washington, D.C.: World Bank Publications.

National Assembly of Vietnam. (2003). *Land law (in Vietnamese)*. Retrieved from http://www.chinhphu.vn/portal/page/portal/chinhphu/hethongvanban?class_id=1&mode=detail&document_id=32479.

National Assembly of Vietnam. (2005). *Tendering law (in Vietnamese)*. Retrieved from http://vanban.chinhphu.vn/portal/page/portal/chinhphu/hethongvanban?class_id=1&mode=detail&document_id=29535.

Nguyen, X. T., & Dapice, D. (2009). *Vietnam's infrastructure constraints* (Policy Dialogue Paper No. 3). Cambridge, MA: ASH Institute for Democratic Governance and Innovation John F. Kennedy School of Government.

Ofori, G. (1990). *The construction industry: Aspects of its economics and management*. Singapore: NUS Press.

Ofori, G. (2000). *Challenges of construction industries in developing countries: Lessons from various countries*. Paper presented at the 2nd international conference on construction in developing countries. Retrieved from http://rec.hku.hk/steve/Msc/08-09/reco6005/RFF_challenges.pdf.

Pincus, J., & Nguyen, T. (2004). *Poverty reduction and strategy process and national development strategies Asia: Country study Vietnam*. London: University of London.

Priemus, H. (2010). Mega-projects: Dealing with pitfalls. *European Planning Studies, 18*(7), 1023–1039.

Prime Minister's Office of Vietnam. (2009a). *Decree No. 69/2009/NĐ-CP: Additional regulations on land use planning, land price, land acquisition, assistance and resettlement (in Vietnamese)*. Retrieved from http://www.moj.gov.vn/vbpq/Lists/Vn%20bn%20php%20lut/View_Detail.aspx?ItemID=24276.

Prime Minister's Office of Vietnam. (2009b). *Decree No. 112/2009/NĐ-CP: On management of project costs (in Vietnamese)*. Retrieved from http://vanban.chinhphu.vn/portal/page/portal/chinhphu/hethongvanban?class_id=1&mode=detail&document_id=92178.

Quang, N., & Kammeier, H. D. (2002). Changes in the political economy of Vietnam and their impacts on the built environment of Hanoi. *Cities, 19*(6), 373–388.

Ringle, C. M., Wende, S., & Will, A. (2010). Finite mixture partial least squares analysis: Methodology and numerical examples. In V. E. Vinzi, W. W. Chin, J. Henseler, & H. Wang (Eds.), *Handbook of partial least squares* (pp. 195–218). Berlin: Springer.

Sohail, M., & Cavill, S. (2008). Accountability to prevent corruption in construction projects. *Journal of Construction Engineering and Management, 134*(9), 729–738.

Tan, W. C. K. (2007). *Practical research methods* (3rd ed.). Singapore: Prentice Hall.

Thanh, N. (2012). Lengthy land disputes partially caused by problems with legal setting (in Vietnamese). *Vietnam Statues Online*. Retrieved from http://www.phapluatvn.vn/thoi-su/toan-canh/201211/Khieu-kien-dat-dai-keo-dai-mot-phan-do-van-ban-luat-2072457/.

Thinh, L. V. (2011). *Cracks on the surface of Thang long bridge: The underlying cause*. Vietnam: The State Authority for Construction Quality Inspection of Vietnam.

Tinh, D. (2012). Settlement and crack in Cau Gie–Ninh Binh high way: An impact of time pressure. *Vietnam Economic Times*. Retrieved from http://vneconomy.vn/20121202114022183P0C9920/cao-toc-cau-gie-ninh-binh-lun-nut-hau-qua-ep-tien-do.htm.

Todaro, M., & Smith, S. (2009). *Economic development* (10th ed.). Boston: Addison Wesley.

Transport Engineering Construction, & Quality Management Bureau. (2011). *Overview of building quality management in transport infrastructure projects (in Vietnamese)*. Vietnam: Ministry of Transport of Vietnam.

Van Wee, B. (2007). Large infrastructure projects: A review of the quality of demand forecasts and cost estimations. *Environment and Planning B: Planning and Design, 34*(4), 611.

Chapter 9
Conclusions

Abstract This chapter summarizes the subject matters of this study and major research findings. More importantly, the chapter presents research recommendations. Recommendations are prioritized based on the extent to which a particular factor acts as a constraint to the efficient use of public investments in infrastructure in Vietnam, including measures to sustain governmental commitment and to improve institutional capacity, to enhance transparency and accountability in infrastructure development, to improve the land acquisition framework, and to strengthen building capacity of local firms. The chapter ends with research contributions, limitations and recommendations for future works.

Keywords Efficient use of infrastructure investment · Influencing factors · Capacity building · Policy implications · Long-term growth

9.1 Introduction

Infrastructure funding has been shown to be critical to the questions of trade and economic growth (see Chaps. 3 and 4). This research examines the question of efficiency in using public investments in infrastructure to support trade and economic growth in the context of Vietnam. Over the last decade, Vietnam has invested more than 7.5 % of GDP in electricity supply, transport and communications (see Table 8.1). Over the next 5 years, Vietnam will need about US$139 billion to be invested in infrastructure for promoting trade and economic growth (Moore et al. 2010). Apparently, whether the amount of US$139 billion set aside for infrastructure development would make a long-term contribution to improving the nation's public infrastructure depends on the quality of the government's infrastructure plan and delivery. More importantly, what factors influence how well the nation's fixed assets are planned and built?

The first objective of this study is to review how public investments in infrastructure are used to support trade and economic growth in the literature. The study

G. Dang and L. Sui Pheng, *Infrastructure Investments in Developing Economies*,
DOI 10.1007/978-981-287-248-7_9

reviews the literatures on construction and economic development, international trade theories and infrastructure development in Chaps. 3, 4 and 5. The literature review suggests that public investments in infrastructure play an important role in supporting trade and economic growth by providing physical structures and facilities for the economy as a whole and improving access to international markets. The second objective of this study is to examine how infrastructure is planned and built in Vietnam. The objective has been achieved in Chap. 6. Infrastructure development, government planning and related issues in Vietnam have been reviewed in the chapter.

In achieving the last three objectives, this study examined how efficiently public investments in infrastructure used over the last decade support trade and economic growth in Vietnam. With this backdrop, the study tested a model that examined the impact of institutional capacity of the government, forecasting capacity, transparency and accountability, land acquisition and capacity of local construction firms on the efficient use of public investments in infrastructure in Chap. 8. The following findings and recommendations emerged from the data analyses.

9.2 Major Findings

9.2.1 Efficiency Level of Public Investments in Infrastructure

The efficiency of public investments in infrastructure is measured by the extent to which the public investments in infrastructure are well used for improving the quality and the extensiveness of infrastructure networks for trading expansion, and subsequently economic growth. Public investments in infrastructure are measured by state investments (as % of GDP) in electricity supply, transport and communications. The data were collected from the various Statistical Yearbooks of Vietnam issued by the General Statistics Office of Vietnam. The quality and extensiveness of infrastructure networks for promoting economic activity are measured by the competitiveness index of Vietnam in terms of infrastructure development. The index was collected from the various Global Competitiveness Reports published by the World Economic Forum.

The results of the longitudinal survey show that from 1999 to 2010, the efficiency of public investments in infrastructure increased in some years, but decreased in others. More importantly, the marginal effect of the public investments in infrastructure (the impact on efficiency of one-unit change in public investments in infrastructure) was an inverted U-shaped. Since 2002, the marginal effect has been declining. This reflects the falling trend in efficiency of public investments in infrastructure for supporting trade and economic growth in Vietnam.

9.2.2 Factors Affecting the Efficient Use of Public Investments in Infrastructure

From the literature review, 35 problems (see Table 8.4) were identified. These were used in a questionnaire survey to explore what problems would significantly affect the efficient use of public investments in infrastructure in Vietnam. The questionnaire was sent to both the government officials and the major developers, financial institutions, traders, contractors and consultants involved in planning and delivering infrastructure, as well as in utilizing infrastructure services for trading activities. Spearman ranking correlation test shows that there is a high degree of agreement between the two groups on the level of occurrence and criticality of problems. From the simple data analysis, the top 27 ranking problems in terms of the median critical ratings were selected for further factor analysis.

Factor analysis was employed to capture the multivariate relationship existing among the problems. The Bartlett test of sphericity and the measure of sampling adequacy (MSA) justifying the application of factor analysis suggest that a structure exists to group problems. Based on the procedure in applying factor analysis (Hair et al. 2010), 25 problems were retained and structured into five distinct group factors, including insufficient institutional capacity of the government, lack of transparency and accountability in infrastructure development, lack of an effective land acquisition framework, inadequate forecasting capacity and insufficient building capacity of local firms.

Based on the results of factor analysis, a model of major group factors affecting the efficient use of public investments in infrastructure was developed. In the model, lack of transparency and accountability in infrastructure development, lack of an effective land acquisition framework, inadequate forecasting capacity and insufficient building capacity of local firms were all positively related to the inefficient use of public investments in infrastructure. In addition, insufficient institutional capacity of the government was considered as the root construct having critical effects on other factors, and ultimately on the inefficient use of public investments in infrastructure in Vietnam. The assessment of the theoretical model was conducted with PLS-SEM. PLS-SEM assessment consists of two parts: evaluation of measurement model and evaluation of structural model. Results of the measurement model suggest that reliability and validity of measured indicators (construct reliability, convergent validity and discriminant validity) are supported (see Sect. 8.4.2).

Evaluation of the structural model was conducted using the coefficient of determination R^2, the assessment of the significance of path coefficients and the predictive relevance of the structural model (see Sect. 8.4.3). The results suggested that all relationships in the model are statistically significant and robust to PLS-SEM techniques, except for the impact of inadequate forecasting capacity. The theoretical model is thus generally supported. Each relationship is symptomatic of a set of underlying causes. The results were discussed together with in-depth interviews. The following are the main results of the analysis.

9.2.2.1 Effects of Inadequate Forecasting Capacity

The analysis shows that inadequate forecasting capacity has no statistically significant effects on the efficiency of public investments in infrastructure in Vietnam. Issues with forecasting capacity have previously been associated with forecasting inaccuracy (Ascher 1979; Flyvbjerg et al. 2002, 2005). Others argued that misleading forecasts are not truly caused by inadequacies in forecasting capacity (Flyvbjerg 2007). The results of the empirical analysis in this research suggest that inadequacies in forecasting capacity per se do not matter. Although improvements in forecasting capacity have been made in Vietnam, a substantial gap between forecasts and the actual results of infrastructure development planning still exists (Nguyen and Dapice 2009). This points to the importance of examining other factors that would affect efficiency of public investments in infrastructure in Vietnam.

9.2.2.2 Effects of Lack of Transparency and Accountability in Infrastructure Development

The analysis shows that lack of transparency and accountability in infrastructure development is largely responsible for the inefficient use of public investments in infrastructure in Vietnam. The low level of transparency and accountability poses significant risks of corruption, and fraudulent and law-breaking practices of government officials and local firms in public projects. This can have serious detrimental effects on the quality of infrastructure development and ultimately on the efficiency of public investments in infrastructure. There are three broad areas of problems associated with transparency and accountability in currently funded infrastructure development:

(i) Lack of transparency and competitiveness in bidding and award: the inadequate consideration of appropriate qualifications when selecting firms, the absence of measurable criteria for capturing the difference in qualifications of tendering firms and the lack of a mechanism to verify the tendering applications of firms; the avoidance of open competitive bidding by adjusting the contract valuation below a certain level or dividing procurement into smaller contracts.

(ii) Complicated and bureaucratic administrative system in infrastructure development: complicated rules and procedures for review and approval across levels of government; and weak accountability for government officials' actions and decisions and a culture of collective responsibility among government officials.

(iii) Poor enforcement of the existing legislation and regulations relating to corruption: the combination of many roles of the government in infrastructure development; lack of effective protection for whistle-blowers and punishments stringent enough to prevent corruption.

9.2.2.3 Effects of Lack of an Effective Land Acquisition Framework

The analysis shows that lack of an effective land acquisition framework is a statistically significant cause of the inefficient use of public investments in infrastructure. Lack of an effective land acquisition framework can lead to controversies over compensation packages and increasing claims and disputes in land acquisition. The empirical analysis in this study reveals that these difficulties in land acquisition can slow down the project development, causing significant delays and cost overruns. There are two broad areas of difficulties with the current land acquisition framework in Vietnam:

(i) Bureaucracy in land disputes and claim settlements: bureaucracy in issuing land-use certificates; high tax and legal scrutiny required for transfer of land-use titles; and overlapping and inconsistency in land-use regulations.
(ii) Inadequate compensation for land acquired for public projects: large gaps between state price frameworks and market prices of land, the loss of business of affected land users, inadequate assistance for resettlement, inadequate public financial resources for compensation, speculative purchases of land surrounding the area to be acquired; and
(iii) Lack of a clear spatial plan for infrastructure development.

9.2.2.4 Effects of Insufficient Building Capacity of Local Firms

Data analysis in this study reveals that there is a certain relationship between building capacity of local firms and the efficient use of public investments in infrastructure in Vietnam. Local firms lack the expertise and resources to guarantee the quality and professional competence of the workforce in large infrastructure projects. Consequently, these constraints result in the deterioration of efficiency of funded infrastructure development. The relationship suggests that increasing the building capacity of local firms can raise the efficiency of funded infrastructure development, benefiting both the government and local firms. Thus, increasing the building capacity of local firms is an important part of the national infrastructure development plan. The major challenges with the current building capacity of local firms in undertaking public projects include:

(i) With state-owned companies or PMUs assigned to be responsible for project development: lack of managerial expertise for large infrastructure projects, lack of thrust on optimization of costs, low level of accountability, inadequate and slow release of funds and payments.
(ii) With supervision consultants: low consultant fees, low level of accountability and professionalism, lack of systematic training, certification and registration.
(iii) With designers and engineers: lack of experienced and highly qualified designers and engineers, lack of thrust on optimization of design,

inadequate topographical surveys; lack of resources and capacity for quality control, low level of accountability and professionalism.

(iv) With contractors: lack of capable contractors for handling large infrastructure projects; lack of experienced staff and skilled labour, lack of a quality management structure and culture, and financial and modern equipment constraints.

9.2.2.5 Effects of Insufficient Institutional Capacity of the Government

A result that should be of particular concern to the government is the detrimental effects of insufficient institutional capacity of the government. The empirical results show that insufficient institutional capacity of the government has produced detrimental effects far beyond other factors. As insufficient institutional capacity of the government influence forecasting, transparency and accountability, land acquisition and capacity of local construction firms, their effects spill over into the efficiency of public investments in infrastructure. Insufficient institutional capacity of the government restricts government's ability to generate good planning and sensible policies, as well as the ability to create a framework that ensure the good performance of various actors in carrying out central policies. There are three broad areas of difficulties with the current institutional capacity of the government, including:

(i) Inadequate coordination in infrastructure planning: inadequate coordination between sectoral ministries and between lower levels of government; lack of incorporation of financial considerations into infrastructure planning.

(ii) Poorly executed decision-making: lack of national interests and lack of a long-range vision on the part of political leaders and government bureaucrats, lack of competent and qualified planners, lack of ex-ante assessment by independent-review bodies and scientific and professional community, lack of an independent mechanism for monitoring decisions and actions made by government officials.

(iii) Deficient organizing for delivery: the absence of a high-powered agency responsible for the entire delivery of national infrastructure programs, the lack of managerial capacity of local governments in infrastructure development, the absence of measurable criteria for evaluating (physical and financial) performance and poor ex-post assessment to verify whether infrastructure projects meet objectives, and changing and inconsistent laws and regulations in infrastructure construction (including construction investments and financing, and bidding regulations).

Findings of the research identified above are summarized in the framework illustrated in Fig. 9.1.

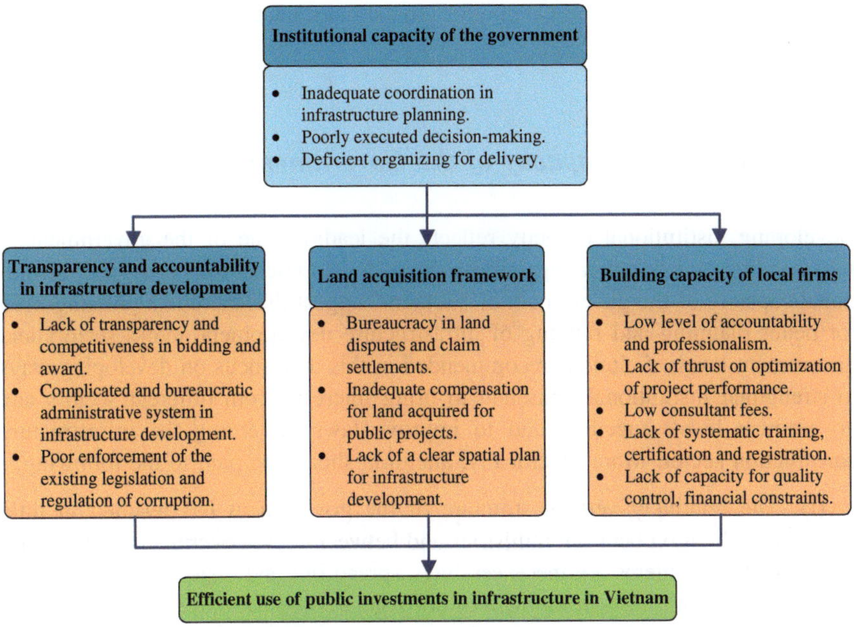

Fig. 9.1 Major factors affecting the efficient use of public investments in infrastructure in Vietnam

9.3 Recommendations

From the factor analysis, four groups of factors emerge as key challenges in moving towards the efficient use of public investments in infrastructure. The following recommendations stem from the underlying causes of each factor and suggestions discussed in Sect. 8.5. Recommendations can be prioritized based on the extent to which a particular factor acts as a constraint to the efficient use of public investments in infrastructure. Results of the factor analysis suggest that the lack of institutional capacity of the government is the root construct. Transparency and accountability have the second biggest impact, and the next two factors are land acquisition and building capacity of local firms (see Fig. 8.3). As a result, the starting point for recommendations would be solutions to sustain governmental commitment and to improve institutional capacity. Other recommendations are measures to enhance transparency and accountability in infrastructure development, to improve the land acquisition framework and to strengthen building capacity of local firms. Recommendations would mainly require capacity building or changes in policies and regulations. While capacity building needs a medium-term time frame, changes in policies and regulations can be implemented in a shorter time frame. The central government, including the Ministry of Planning and Investment, Ministry of Finance, Ministry of Construction, Ministry of Industry and Trade, Ministry of Natural Resources and Environment, Ministry of Transport and

Ministry of Information and Communication should, take a lead in organizing and directing these efforts.

9.3.1 Institutional Capacity of the Government

Developing institutional capacity reflects the leading role of the government in spearheading and coordinating infrastructure development. Concerns over institutional capacity of the government of Vietnam suggest that there is a pressing need for better planning and running of infrastructure development across sectors and government levels. Solutions recommended in this area focus on developing ways to strengthen institutional links and to bring interests from different sectors and government levels together, and to improve laws governing the infrastructure construction sector. More specifically, attention should be paid to the following:

(i) The relatively vertical structure of the government has resulted in little coordination between ministries and between local governments. Outcomes of the problems are the excessively spread out and disjointed investments. As a result, the development of horizontal coordination channels across the ministries and across the local governments that have infrastructure responsibilities would help strengthen their coordination and cooperation.

(ii) The mismatch between budget allocation plans and investment plans requires the government to develop a process for integrating budgeting decisions with investment decisions made by ministries and local governments, and ensuring that projects are to be approved with corresponding plans for budget allocation.

(iii) With reference to decision-making, there is a need for safeguarding national interests. The underlying causes suggest that there is a strong need for an independent advisory board within the central government composed of a wide range of stakeholders from government officials, academics and independent consultants to monitor the direction of infrastructure development and develop tools to ensure that sectoral and local interests are aligned with national priorities. The government should also focus on building a core cadre of planners within the central government that have the ability to generate a strategic vision for future infrastructure systems.

(iv) The current decentralization arrangement in the delivery of national infrastructure requires the establishment of a high-level government body (within the central government) that has the resources and the mandate to coordinate, monitor and resolve any disputes between lower levels. The concerns over the evaluation system of delivery performance suggest the need to introduce a system of regular evaluation with (physical and financial) performance indicators that can measure progress towards goals and the real costs of infrastructure delivery, and ensure that public disclosure (as a driver for more accurate evaluation) is provided, thus providing actionable advices for related government agencies to respond to problems.

(v) The laws governing the construction industry is another institutional barrier that requires the government to conduct a regulatory review of the current construction laws and regulations, especially those related to construction investments and financing, and bidding. Findings of the review would be used as a basis for building a unified legal framework.

9.3.2 Transparency and Accountability

Transparency and accountability are primary concerns for the efficient use of public investments in infrastructure. Based on major weaknesses in transparency and accountability identified in infrastructure development, actions centre on competition in bidding, administrative reforms and explicit anti-corruption measures are suggested.

(i) Supporting greater use of competitive bidding for infrastructure projects, the government should revise the Tendering Law to ensure that firms are selected based on both their qualifications and prices, rather than focusing on tender price only. Accordingly, there is a need for introducing measurable criteria and procedures for capturing and evaluating the difference in the qualifications of tendering firms necessary to ensure their capability to fulfil the contract. There is also a need to set up a mechanism to verify their tendering applications and to prevent fraud. In order to increase competition in bidding, the government should also develop tendering regulations that forbid the selection of valuation methods and division of procurement with the view to avoid open-tendering.

(ii) Dealing with the bureaucracy in infrastructure development requires the establishment of a system where lower level governments are capable of making decisions without referring back to higher level governments or the Prime Minister. For such a system to work, training for local government officials should be offered to enhance their management capability relating to infrastructure development. There is also an issue concerning how government officials should be held accountable for their actions and decisions. The low level of accountability and the culture of collective responsibility suggest more needs to be done in improving information disclosure concerning the decisions and actions of government bureaucrats, subjecting their conduct to public scrutiny; and establishing a system of rewarding good practices and punishing bad ones.

(iii) The combination of many roles of the government suggests that there is a need for establishing an independent regulatory institution (independent of ministerial influence), and the use of external auditors to detect fraud and corruption, and enforce rules when required. In addition, providing effective measures to ensure the confidentiality of whistle-blowers, and stricter punishments for offenders are also suggested to help prevent corruption.

9.3.3 Land Acquisition

Increasing demand for infrastructure would inevitably involve considerable instances of land acquisition. Difficulties in land acquisition in Vietnam show that the key is to ensure that adequate compensation is provided for affected land users. Recommendations are:

(i) The starting point would be addressing bureaucracy in issuing land-use certificates, which give land users the right to claim for compensation for the land acquired. The current bureaucratic process suggests that simplification of land registration arrangements would be particularly important. Furthermore, measures should be studied to reduce land-use tax burden for land users once land-use certificates are issued. The current regulations for dispute resolution should be reviewed to ensure that land disputes and claims are fairly and consistently settled.

(ii) Under the current mechanism in public projects where the government is both the valuer and the takeover institution, there is no room for those displaced to challenge the compensation rates set by the government. Developing a mechanism for land users to discuss/negotiate compensation rates with relevant authorities would thus help fill the gap between the state price framework and market prices of land. In addition, the government should develop guidelines and criteria for building compensation packages that recognize loss of business of affected land users.

(iii) Although assistance measures for resettlement are required by Land Law, in practice these are not adequately implemented. Improving transparency and accountability in practice is thus necessary. The starting point would be to develop an independent monitoring system for the implementation of compensation and resettlement.

(iv) Paying proper compensation is becoming more problematic due to the high price of land. A possible direction suggested in this study would be for the government to develop land taxes to capture a proportion of the increase in land value associated with infrastructure improvements. The collected taxes would not only serve as an avenue to increase revenue for compensation, but also have the happy side-effect of reducing social inequality and gaining public support.

(v) Speculative purchases of land remain the major force that push up land prices artificially. Addressing the issue is thus particularly important to improve the efficiency of land markets, and to reduce compensation costs. A suggested solution is to establish procedures and criteria for the transparent and competitive sale of land in the area to be acquired for infrastructure development.

(vi) There is also a suggestion to develop a clear spatial plan for infrastructure development, so that land would be set aside in advance for infrastructure development, and resettlement costs would be significantly reduced.

9.3.4 Building Capacity of Local Firms

Building capacity of local firms is identified as an essential element for achieving the efficient use of public investments in infrastructure. There is room for improvement in many aspects of building capacity of local firms. This includes improving management capability, developing human resources, increasing level of accountability and professionalism, and strengthening financial and equipment rental markets. The details of recommendations are:

(i) For state-owned companies or PMUs assigned to be responsible for project development:

- Since there are no rules for assigning state-owned companies or PMUs, the government should set up the procedures for appointing PMUs, and providing criteria for selecting capable PMUs. Training for PMUs should be offered to enhance their management capability, and capacity for quality control, contract and contractor management.
- To encourage greater accountability, the government should also provide rules for PMUs to manage project effectively, make strategic decisions about the project performance and hold the project manager accountable for its performance. Selected PMUs should be tied to financial responsibility related to costs of construction, operations as well as maintenance.
- In order to accelerate the release of funds and payments and reduce uncertainties over budget allocation, the government should simplify the release procedures and match medium and long-term project financing needs with the government budget allocation plans.

(ii) For supervision consultants:

- More attention should be given to promote the supervision consultancy profession. National government agencies and the industry associations should take a lead in improving consultant fees for supervision consultants and providing incentives to retain people in the sector.
- In order to achieve greater accountability and professionalism, a formal system of performance evaluation should be established. In addition, training, certification and registration should also be systematically introduced to enhance the quality of human resources.

(iii) For designers and engineers:

- A long-term capacity development framework for local consultants should be introduced to address the needs for high-quality designers and engineers.
- Government agencies should introduce merit-based rewards or other incentive schemes to encourage designers and engineers to optimize design. Also, there should be a formal system of performance evaluation and forms of liability on the consultants on the costs incurred due to poor-quality of design. Training should also be offered to enhance their capacity for quality control.
- The government agencies should develop measures to ensure that adequate funds are set aside for topographical surveys; and set standards to ensure the quality of survey reports.

(iv) For contractors:

- The government and the industry should focus on ways and means to facilitate and attract larger contractors. For small-scale contractors, training should be offered to enhance their management and financial management capability, and capacity for quality control. Contractors should set up their in-house quality management systems rather than to rely on other parties.
- The government and the industry should provide incentives or create an enabling environment for banks and other financial institutions to develop financial markets that provide commercial loans for small-scale contractors. Likewise, measures to foster an equipment rental market should be instituted to meet the demand for modern equipment from small contractors.

9.4 Validation

The qualitative validation exercise for this research study was conducted in April 2013. Five experts in Vietnam were selected for the validation interviews. As explained in Sect. 7.7, they are senior stakeholders with one involved in the public sector, two in academics and two in the industry. They have been involved with the process of public investments in infrastructure since the mid-1990s. Major steps in the validation included presenting these experts the research findings and asking them to give comments on the reliability of the research findings and the relevance and practicability of the research recommendations. All experts commented that the research findings were comprehensive as they reflected on the current situation in Vietnam. With reference to the research recommendations, the interviewees commented that they were generally relevant and practical. Two experts said that the research recommendations are comprehensive and practical. One expert said that many of the research recommendations are outstanding. One expert said the research recommendations can contribute to improve the efficiency of the use of public investments in infrastructure in Vietnam. One interviewee was not impressed with the research recommendations. However, when the expert was asked for suggestions on how to improve them, the expert could not give a clear answer. In addition, all experts emphasized that Vietnam has made a number of changes in laws and regulations to close the gaps between international principles and Vietnamese policies on paper at least. However, difficulties remain in the execution of these laws and regulations. Therefore, the experts suggested that recommendations for capacity building should receive more attention. The highlights of the validation interviews are presented below.

9.4.1 Institutional Capacity of the Government

All experts emphasized that in order to avoid excessively spread and disjointed infrastructure investments, identifying strategic goals for infrastructure development is crucial. Identifying strategic goals for infrastructure development should be based on a comprehensive study to select priorities that could seize the opportunities and utilize resources on a few strategic goals. The experts agreed with the research findings and recommendations as explained by one expert:

> Functions and roles of the government at different levels (vertically and horizontally) and at various stages in policy making and planning have to be more clearly defined; and budget must be adequately allocated in accordance with investment schedule. Roles of "relevant authorities" must be clearly defined since in many cases, interference of relevant authorities in various stages caused waste and created obstacles in infrastructure development.

Although one expert disagreed with the research finding that there is a lack of national interests among government officials, the expert's explanation is in agreement with the research finding:

> There is not a lack of national interests among government officials. The problem is that they abuse public funds to serve their local interests.

Two experts suggested that the government should use professional associations as independent advisory boards as explained by one of the experts:

> Professional associations can serve as independent advisory boards, helping to monitor the direction of infrastructure development and ensure that sectoral and local interests are in line with national priorities. Besides, the State (the Party, National Assembly and Government) can consult with professional associations when preparing master plan for infrastructure development and outlining strategic vision of a future infrastructure system.

Given the role of professional associations in Vietnam, the suggestion is appropriate and can be added to the research recommendation.

One expert suggested that infrastructure should be developed systematically at local level and national level, and in collaboration with neighbouring countries for long-term development goals. Since the study focuses on national level, the suggestion of regional collaboration in infrastructure development can be considered in future research work.

The experts concurred with the research findings and recommendations of the establishment of a high-level government body for coordination in the delivery of national infrastructure programmes. One expert emphasized that:

> The establishment of a high-level government body is necessary for coordination in delivery of national infrastructure programs. The government body should regularly use the evaluation system with performance indicators comparing planned goals with actual progress and costs of infrastructure delivery; and publicize these results (as a driver for more accurate evaluation), thus providing actionable advices for related government agencies to respond to problems.

Regarding the research recommendation for a formal monitoring and evaluation system, one expert highlighted that:

> Currently lower level governments are required to periodically report performance of public investments in infrastructure to higher levels. However, the lack of specific guidelines with measurable criteria for performance evaluation reduces the effectiveness of the reporting system. It is necessary to issue standard principles for performance evaluation of public investments in infrastructure.

Relating to changes in laws and regulations, one expert commented that the government is taking steps to review laws and regulations governing the construction industry, and amendments are expected to be made to address the inconsistencies among them.

9.4.2 Transparency and Accountability

The experts shared the same opinion with the research findings that lack of transparency and accountability in infrastructure development is one of the pressing issues in Vietnam. The experts emphasized that existing laws and regulations have provided a set of rules to ensure the good performance of various actors in carrying out central policies. The experts agreed with the research findings that the poor execution of the existing laws and regulations, and the absence of a set of measurable criteria for performance evaluation of the actors involved pose a significant threat to the efficient use of public investments in infrastructure in Vietnam. With reference to bidding, one expert explained:

> Although the Tendering Law mentioned the acceptance of the lowest evaluated tender bid, the current Construction Law allows firms meeting minimum requirements of relevant qualifications be selected on the basis of rational tender price, which might not necessarily be the lowest one. However, in practice as long as firms meet minimum requirements of relevant qualifications, firm selection focuses on tender price, which can be the lowest. The major issue is that there are no measurable criteria to capture the difference in qualifications that go beyond the minimum requirements. With reference to a mechanism to verify firms' tendering applications, the current Tendering Law requires that each province will select a top leader (at least vice president of provincial people's committee) to be in charge of

> tendering at the provincial level. The selected government official has the right to cancel the tendering and reject the tendering results if any fraudulent practices are discovered. However, the absence of guidelines and measurable criteria for verification has made it difficult to prevent fraud. Besides, tendering applications with subcontracting work is allowed in the current Tendering Law. However, there is no mechanism to verify the relevant qualifications of the subcontractors.

One expert added that there should be measures to prevent firms affiliated with related government ministries or procurement officials from bidding. The expert stressed the research recommendation of the use of independent monitoring institutions, and external auditors throughout various stages of infrastructure development to deal with the combination of many roles of the government.

Relating to decentralization in infrastructure development, one expert emphasized on the research findings that:

> Decentralization is necessary, but training should be provided for lower government levels. The government should develop specific regulations for lower government levels on how to monitor and evaluate infrastructure development. Financial liability should be applied for fraudulent practices made by government officials.

9.4.3 Land Acquisition

Regarding land acquisition, one expert commented that the research recommendations of measures to ensure adequate compensation to be provided for affected land users are novel. The expert also concurred with the research recommendations that establishment of an independent monitoring system for the implementation of compensation and resettlement is necessary to improve transparency and accountability in land acquisition.

One expert focused on the research finding that inconsistency in current regulations for land use and dispute resolutions should be addressed. In addition, the expert proposed that the vicinity of the infrastructure site should be acquired together with the infrastructure site. The local government can sell the vicinity and create equality between those displaced and those benefiting from infrastructure improvements. The suggestion that has been discussed in Chap. 8 might not be relevant in urban areas.

Two experts are, however, concerned with the difference in compensation rates offered across adjacent land areas and thus suggested some solutions. As explained by one of the experts:

> Land-related policies should be consistent for at least 5 years. According to current Land Law, compensation rates vary between land zones. As a result, land users in different land zones would be offered different compensation rates, even though they stay in adjacent land zones. The mechanism creates social inequality among land users in adjacent areas. To deal with the issue, there should be special compensation valuation methods for adjacent land zones and land areas where land users can benefit from infrastructure improvements. There should also be regulations to prevent fraudulent practices in order to get higher compensation rates, delaying land acquisition.

One expert emphasized that the research recommendation for the development of a mechanism for land users to negotiate compensation rates with developers in public projects would be the answer to the dissatisfaction over land compensation rates among land users, especially those in adjacent areas.

9.4.4 Building Capacity of Local Firms

With reference to building capacity of local firms, one expert highlighted that the two most pressing issues in infrastructure delivery in Vietnam are quality and time and that improving capacities of related parties are important to address these issues. The expert suggested that capacity building for local firms should focus on building high-quality firms for niche markets.

Regarding the capacity of PMUs, one expert emphasized and concurred with the research finding that PMUs representing the relevant ministries for project development must have sufficient capability and are allowed to make decisions independently of the ministries in the areas where the PMUs are capable of. One expert suggested that the government should encourage contracting out project management work through bidding rather than performing them in-house to deal with the lack of managerial expertise of the PMUs and to improve the level of professionalism. The suggestion can be added to the list of research recommendations.

Regarding the capacity for supervision consultants, one expert focused on the research recommendation that training, certification and registration should be renewed to improve the level of professionalism and work ethics. The expert also recommended that there should be regulations to revoke the professional certificates of those found to have fraudulent practices. The suggestion is suitable to improve the level of accountability of supervision consultants.

Relating to the capacity of designers and engineers, one expert emphasized the need to improve the level of the accountability. The expert agreed with the research recommendation of setting up a formal system of performance evaluation, suggesting that:

> Their performance should be evaluated through a set of criteria. Performance should be evaluated through aesthetics, functional, sustainable and economical criteria of designed buildings and structures. In addition, information on capacity of designers and engineers should be posted online and updated annually.

For contractors, the expert also suggested that:

> Firms should be classified according to their capacity. Measurable criteria for firm classification should be formulated. List of classified firms should be posted online and updated annually.

The suggestions are appropriate to create an incentive for firms to improve their capacities.

Relating to the research recommendation of introducing incentive schemes to improve the level of professionalism among supervision consultants, designers and engineers, one expert suggested that there should be regulations to increase consultant fees. The expert stressed that the current low level of consultant fees could pose a high risk of corruption and fraudulent practices.

Out of these discussions, a few suggestions made by these experts can be added to the research recommendations. The validated research recommendations are finally summarized in Fig. 9.2.

9.5 Contributions and Implications

9.5.1 Theoretical Contributions

The current research has several contributions to knowledge. The study addressed the main research question of what factors influence/limit the efficient use of public investments in infrastructure to support trade and economic development.

The study explored public investments in infrastructure based on an entire management system that encompasses planning, budgeting, inter-ministry and inter-local government cooperation, and delivery. The approach focuses on the macro-level in order to identify the sources of issues in using public investments in infrastructure to support trade and economic development across government departments and policy spheres.

Employing the above-mentioned approach, the research supplements the understanding of the role of government in the theory of coordination failure in economic development. The theory of coordination failures argues that when complementaries exist, but are not seen to be feasible from individual private investors' point of view, the active roles of governments as coordinators and facilitators are needed (Rosenstein-Rodan 1943). The focus of the policy recommended is thus the use of a government-led approach/investment as a basis for coordinated efforts (Krugman 1995). Similar to other developing countries during their early stage of development, massive injections of funds in infrastructure development have become the keystone in the strategy for rapid economic growth in Vietnam. Meanwhile, the potential role of government in ensuring productive deployment of public investments in infrastructure has been downplayed. Findings of the study suggest that as the country transforms into a middle-income country, the potential role of the government extend beyond provision of capital. Massive funding is necessary for accelerated economic development, but it is not sufficient to ensure the long-term benefits. The way that the government plans and runs infrastructure policies lies at the root of the efficient use of public investments in infrastructure. Infrastructure interventions in a middle-income country are likely to require the government to take the lead in ensuring that development funds induce a long-term and sustainable growth. State capacities are thus critical to support the efficient use of public investments in infrastructure.

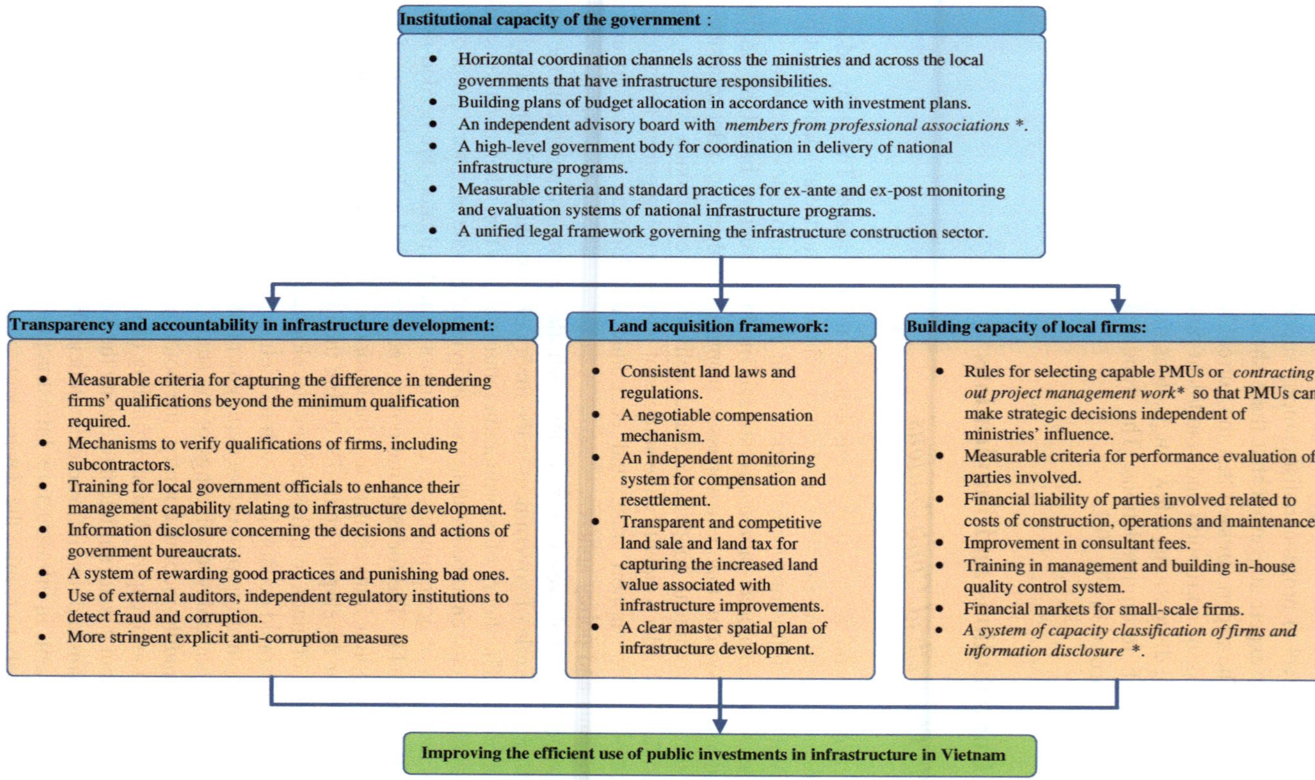

Fig. 9.2 Framework for developing capacities for the efficient use of public investments in infrastructure in Vietnam. *suggested by experts during validation exercise

In addition, the research supplements the theory of coordination failure by highlighting the linkages between infrastructure investments for trade and economic development and infrastructure delivery. Recent works in the literature on coordination failure emphasized that in an economy of a complex system of activities, dealing with the coordination problem requires the government to figure out and identify which combination of activities could bring out a better outcome as a whole (Matsuyama 1997; Hoff 2000). Literature on the linkages between infrastructure and trade as well as the economy focuses on directions in which public investments in infrastructure need to develop for accelerated trade and economic development in developing countries. However, findings of this study show that infrastructure investment is only part of the equation. Impact of infrastructure delivery is just as important. The efficient use of public investments in infrastructure does not end at planning and decision-making of infrastructure investments. There are a number of policies and activities that can cause infrastructure plan disruptions, including budgeting decisions, land acquisition, spatial development, and construction laws and policies. Findings of the research imply that the government should look more into policy initiatives that would significantly improve the management of infrastructure delivery. Well-targeted infrastructure interventions for trade and economic development require infrastructure planning to take into account these policy initiatives, generate strategic vision and develop a mechanism to translate that vision into reality. Understanding of the linkages is critical for devising an effective framework for infrastructure interventions in developing countries. Major theoretical contributions of the study are illustrated in Fig. 9.3.

9.5.2 Practical Policy Implications

Since the liberation process started in Vietnam more than 20 years ago, Vietnam has made an impressive progress in GDP growth (see Fig. 6.1) and rapidly integrated into the world economy. A critical contribution to this success has been the high level of investments in infrastructure. Over the past decade, the government has invested around 8 % of GDP annually in transport, energy, and telecommunications to promote trade and economic growth (see Table 8.1). However, analyses of recent issues in infrastructure development (see Chap. 6) and the marginal effect of public investments in infrastructure over the past decade (see Fig. 8.1) have suggested a falling trend in the efficiency of public investments in infrastructure in Vietnam. Improving the efficiency of public investments in infrastructure is thus necessary for Vietnam to continue and accelerate the growth process. The research offers a consensus framework for developing capacities which are critical for using public investments in infrastructure to support trade and economic development in Vietnam.

Examining how infrastructure investments are formulated points to various coordination problems in national policy-making and the planning process. Findings of the study imply that the government needs a more refined planning system

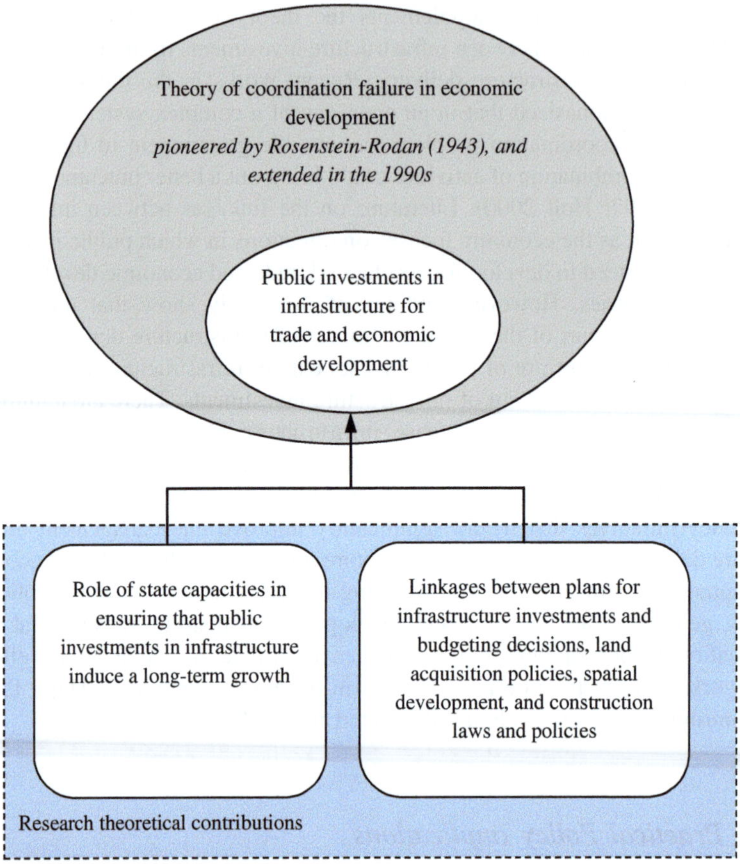

Fig. 9.3 Theoretical contributions of the study

that would bring interests together and solve coordination issues. Thus, the government is recommended to take measures to strengthen institutional links and sustaining commitment to national interests, including cooperative planning mechanisms at the central level, independent consultations, ex-ante and ex-post monitoring and evaluation systems of national infrastructure programmes, and a unified legal framework governing the infrastructure construction sector. Infrastructure planning and delivery demand a higher level of coordination across sectors and government levels. However, the traditional vertical models of planning can result in fragmented policy-making of sectoral ministries and local governments. Lack of horizontal coordination channels is found to be a major challenge for government agencies to coordinate policies in Vietnam. Findings of the research also show that decentralization during delivery can pose significant coordination challenges. Decentralization of government in Vietnam has been increasing in recent years. However, local governments often lack a capacity to drive priorities and ensure the overall quality of national infrastructure programmes. Central institutions that can coordinate lower government levels to implement national infrastructure programmes will play a major role. With extensive involvement of the State in the Vietnamese economy, improving the quality of national planning is critical to develop better infrastructure policies, thus sustaining and accelerating the rate of economic growth. The lessons from Vietnam have broad application since the needs for governments to remake institutional links and bring interests together are commonly found in other developing countries, especially those in their early stages of development (Todaro and Smith 2009).

One of the apparent implications of the research is that transparency and accountability are becoming increasingly significant for the efficient use of public investments in infrastructure. Key challenges to transparency and accountability in infrastructure development in Vietnam include the combination of many roles of the government (as decision makers, regulators, supervisors, clients and so on), the complicated and bureaucratic administrative system (in project planning and approval), and low level of competition in bidding and award. While transparency and accountability are not preconditions for infrastructure development, low level of transparency and accountability can undermine the efficient use of public investments in infrastructure to support trade and economic growth in the long term. Findings of the study highlight that Vietnam needs to address transparency and accountability issues for ensuring the efficient use of public investments in infrastructure, thus generating further economic growth. Transparency and accountability can be improved through specific tools and incentives. The government should consider changes at two main levels: the policies and laws that encourage greater competition in bidding, measures that improve information disclosure concerning decisions made by public authorities, and enforcement of rules; and organizational development (including allocation of responsibilities across government tiers to limit excessive bureaucracy, and systematic monitoring and evaluation).

One of the key factors affecting the efficient use of public investments in Vietnam relates to land acquisition. Findings of the research suggest that the current framework could not correspond to the considerable requirements for land acquisition and resettlement. Current compensation rates are highly distorted due to regulatory and procedural constraints, necessitating changes in land laws and policies. Another important factor that can exert influence on the efficient use of public investments in infrastructure is building capacity of local firms. The fact that implementation of project works often suffers from inefficiency and low quality implies that greater effort needs to be made to strengthen the building capacity of local firms. Findings of the study suggest that there should be policy to address these needs at two main levels: organizational development (the respective roles of PMUs, contractors, supervision consultants, designers and engineers; and their management, financial and technological capabilities); and human resource development (training).

The importance of the efficient use of public investments in infrastructure to support trade and economic growth will increase as demand for large infrastructure facilities is higher. Given limited national resources, in its current phase, Vietnam needs to develop capacities that would remove constraints on the efficient use of public investments in infrastructure. Other developing countries at the same developmental stage can benefit from the research findings and recommendations in dealing with similar problems in infrastructure development. In summary, the research has mapped out the general principles of significant capacity building work which are useful for policy implications, including:

- Institutional capacity for spearheading and coordinating infrastructure development is the centrepiece of the efficient use of public investments in infrastructure;
- Transparency and accountability are becoming increasingly significant for the efficient use of public investments in infrastructure;
- Ensuring adequate compensation is the key to solve the land acquisition issues in infrastructure development; and
- Strengthening the building capacity of local firms is essential for achieving the efficient use of public investments in infrastructure.

Both the academic, public and industry communities in developing countries with similar problems can adopt the research findings as guiding principles to bring changes in their current use of public investments in infrastructure, thus supporting trade and economic growth in the long term.

9.6 Limitations of the Research

There are some limitations on the findings of this research:

- The study considers all infrastructure sectors that support trade and economic growth in Vietnam. Thus, the findings are based on analyses of what limits the efficient use of overall public investments in infrastructure. However, each infrastructure sector is different in the way it supports trade and economic growth. Sectors thus would have different priorities that require different strategic approaches.
- Although the background information on the economic development, trade growth and infrastructure construction of Vietnam is based on longitudinal data analysis, the major findings of the study are connected with data collected at a given time to reflect issues in the current use of public investments in infrastructure in Vietnam. A longitudinal design may be useful to further explain the effects of relevant historical, political and economic events on the use of public investments in infrastructure over a period of time.
- Factors affecting the efficient use of public investments in infrastructure examined in this study are limited to internal factors in Vietnam. However, the influence of external factors is of growing importance, as countries become more integrated into the world economy.

9.7 Suggestions for Future Research

The current study has several contributions to knowledge. There remains, however, potential scope for future research.

First, the current study focused on common issues which can be found across infrastructure sectors. A better understanding of the most pressing obstacles encountered by each sector would be helpful to identify sector-specific areas that need urgent actions. Measures and tools would then be better targeted to sectoral priorities. The current study therefore can be extended to include sector-specific analyses for different infrastructure sectors. Future research can examine the efficiency of public investments in infrastructure of each sector, and uncover the specific factors that determine the sectoral efficiency. Review across infrastructure sectors is required to further explain the performance of each sector and to formulate sectoral strategies.

Second, as the economy liberalizes, factors outside the country can have significant influence on international trade, aid and foreign investments. The current study should be extended to determine how external factors influence the use of public investments in infrastructure.

Third, given the rapid processes of trade and regional integration, regional coordination has become an important issue recently. Regional coordination of infrastructure can play an important role in lowering regional infrastructure costs. Future research can examine the effects of regional coordination of infrastructure policies on the efficient use of public investments in infrastructure, whether regional coordination of infrastructure policies help improve the efficiency of delivery of infrastructure services, and what are the difficulties a country faces in regional coordination of infrastructure policies.

Finally, a longitudinal design can be employed to link relevant historical, political and economic events occurring outside the country to the progress of infrastructure development. This understanding is important to shape development plans and relevant policy outcomes in response to these external factors.

References

Ascher, W. (1979). *Forecasting, an appraisal for policy-makers and planners*. Baltimore: Johns Hopkins University Press.

Flyvbjerg, B. (2007). Megaproject policy and planning: Problems causes, cures. *Environment and Planning B: Planning and Design, 34*, 578–597.

Flyvbjerg, B., Holm, M. S., & Buhl, S. (2002). Underestimating costs in public works projects: Error or lie? *Journal of the American Planning Association, 68*(3), 279–295.

Flyvbjerg, B., Skamris Holm, M. K., & Buhl, S. L. (2005). How (in) accurate are demand forecasts in public works projects?: The case of transportation. *Journal of the American Planning Association, 71*(2), 131–146.

Hair, J. F., Black, W. C., & Babin, B. J. (2010). *Multivariate data analysis: A global perspective* (7th ed.). London: Pearson Education.

Hoff, K. (2000). Beyond Rosenstein-Rodan: The modern theory of coordination problems in development. In B. Pleskovic & N. Stern (Eds.), *Annual world bank conference on development economics* (pp. 145–188). Washington, D.C.: World Bank.

Krugman, P. R. (1995). *Development, geography, and economic theory.* Cambridge: MIT Press.

Matsuyama, K. (1997). Economic development as coordination problems. In M. Aoki, H. K. Kim, & M. Okuno-Fujiwara (Eds.), *The role of government in east Asian economic development: Comparative institutional analysis* (pp. 134–160). New York: Oxford University Press.

Moore, C., Nguyen, T. T. H., & Saladini, M. (2010). *Vietnam infrastructure.* Ho Chi Minh City: Italian Trade Commission and Mekong Research.

Nguyen, X. T., & Dapice, D. (2009). *Vietnam's infrastructure constraints* (Policy Dialogue Paper No. 3). Cambridge, MA: ASH Institute for Democratic Governance and Innovation, John F. Kennedy School of Government.

Rosenstein-Rodan, P. N. (1943). Problems of industrialisation of eastern and south-eastern Europe. *The Economic Journal, 53*(210, 211), 202–211.

Todaro, M., & Smith, S. (2009). *Economic development* (10th ed.). Boston: Addison Wesley.

Appendix A
Questionnaire

Survey of Problems Affecting the Efficient Use of Public Investments in Infrastructure to Support Trade and Economic Development in Vietnam

According to many international surveys, the quality of Vietnam's infrastructure has been identified as the biggest hindrance to the growth of the economy in general and the trade sector in particular. This suggests that there is a need for Vietnam to balance infrastructure quality with trade expansion in order to secure further economic growth. However, heavy investments in infrastructure have not effectively created adequate infrastructure quality. The current situation in Vietnam could result from a number of issues in plan formulation and implementation for national infrastructure development plans. Addressing these issues is important to ensure most efficient use of the funding invested in infrastructure in the long run. Therefore, this survey seeks to shed light on the current factors affecting the efficiency of government funding invested in infrastructure to support trade growth and economic development. The survey is divided into three sections:

- Unexpected outcomes to meet objectives of national development infrastructure plans;
- Problems in infrastructure planning; and
- Problems in delivery of infrastructure programs.

The responses will be analysed to identify the current practices, thus obtaining information on the problems known to participants in Vietnam. The findings are valuable for policy makers and planners to develop tools to improve the efficiency of government-funded infrastructure construction for trade and economic growth.

This survey is conducted as part of a Ph.D. study in the Project Management Program, Department of Building, School of Design and Environment, National University of Singapore. All individual answers will remain confidential. The survey results will only be synthesized and reported in aggregated form.

G. Dang and L. Sui Pheng, *Infrastructure Investments in Developing Economies*,
DOI 10.1007/978-981-287-248-7

Background Information

Types of organizations you have been working:

- ☐ Government agencies
- ☐ Research institutions (research institutes, universities, etc.)
- ☐ Clients/Investors
- ☐ Banks
- ☐ Consultants, contractors
- ☐ Suppliers
- ☐ Traders

Positions or/and academic qualifications: ______________________________

Number of years of work experience: ☐ < 5 years ☐ ≥ 5 years

Part I—Failure to Meet Objectives of National Infrastructure Plans

1. Please tick in the relevant boxes to indicate the extent of the occurrence of unexpected outcomes on objectives of national infrastructure planning.	Degree of occurrence				
	Not significant	Somewhat significant	Significant	Very significant	Extremely significant
	1	2	3	4	5
1.1 Poorly selected infrastructure investments					
1.2 Poorly executed and ineffective infrastructure projects					
1.3 Waste of national resources in funded infrastructure projects					

Part II—Problems in Infrastructure Planning

Please tick your answers in the relevant boxes to indicate the extent of the occurrence of problems in the planning process and their degree of influence on the efficiency of government funding invested in infrastructure to support trade growth and economic development.

2. Capacity for estimating and monitoring of rates of return of infrastructure projects	Degree of occurrence					Degree of influence				
	Not significant	Somewhat significant	Significant	Very significant	Extremely significant	Not critical	Somewhat critical	Critical	Very critical	Extremely critical
	1	2	3	4	5	1	2	3	4	5
2.1 Limitations of forecasting methods and appraisal techniques										
2.2 Inadequate data										
2.3 Inherent problems in predicting the future and monetizing external and indirect effects										
2.4 Lack of experienced forecasters										
2.5 Inadequacy in ex post analysis on whether projects meet objectives										

3. Politicized decision making	Degree of occurrence					Degree of influence				
	Not significant	Somewhat significant	Significant	Very significant	Extremely significant	Not critical	Somewhat critical	Critical	Very critical	Extremely critical
	1	2	3	4	5	1	2	3	4	5
3.1 Political leaders and government bureaucrats make decisions in their personal, sectoral or regional interests (e.g. securing political positions or competing for scarce funds) as opposed to national goals										
3.2 Interest group pressure										

4. Transparency and accountability	Degree of occurrence					Degree of influence				
	Not significant	Somewhat significant	Significant	Very significant	Extremely significant	Not critical	Somewhat critical	Critical	Very critical	Extremely critical
	1	2	3	4	5	1	2	3	4	5
4.1 Lack of consultation with the interested or affected individuals or groups about forecasting methods, criteria of project selection and the determination of planning objectives										
4.2 Lack of independent peer reviews and quality checks on forecasts and planning outcomes by independent-review bodies and scientific and professional community										
4.3 Lack of penal systems to enforce penalties on those that deliberately and consistently produce deceptive forecasts										

5. Institutional weaknesses in planning	Degree of occurrence					Degree of influence				
	Not significant	Somewhat significant	Significant	Very significant	Extremely significant	Not critical	Somewhat critical	Critical	Very critical	Extremely critical
	1	2	3	4	5	1	2	3	4	5
5.1 Poor communication between planning agency and the day-to-day decision-making machinery of government										
5.2 Lack of inter-ministerial and intersectoral coordination										
5.3 Lack of interaction between political leaders and planners with non-governmental actors										
5.4 Incompetent and unqualified civil servants										
5.5 Complicated and bureaucratic administrative system										

Part III—Problems in Delivery of Infrastructure Programs

Please tick your answers in the relevant boxes to indicate the extent of the occurrence of problems in the implementation process and their degree of influence on the efficiency of government funding invested in infrastructure to support trade growth and economic development.

6. Political commitment	Degree of occurrence					Degree of influence				
	Not significant	Somewhat significant	Significant	Very significant	Extremely significant	Not critical	Somewhat critical	Critical	Very critical	Extremely critical
	1	2	3	4	5	1	2	3	4	5
6.1 Lack of political will on the part of top leaders and high-level decision makers										
6.2 Political discontinuity (e.g. changes in the political affiliations of the government responsible for the projects)										
6.3 Political inconsistencies at national level and between different tiers of governments (central, regional, local)										
6.4 Lack of a high-powered government institution that provides an effective mechanism for implementing national infrastructure plans										

7. Corruption in infrastructure construction	Degree of occurrence					Degree of influence				
	Not significant	Somewhat significant	Significant	Very significant	Extremely significant	Not critical	Somewhat critical	Critical	Very critical	Extremely critical
	1	2	3	4	5	1	2	3	4	5
7.1 Lack of transparency and competitiveness in bid processes										
7.2 Discretionary power of individual bureaucrats involved in award of contracts										
7.3 Inadequate financial and physical auditing										
7.4 Inadequate capacity of regulatory bodies to enforce regulations										

8. **Local construction firms' capabilities and resources**	Degree of occurrence					Degree of influence				
	Not significant	Somewhat significant	Significant	Very significant	Extremely significant	Not critical	Somewhat critical	Critical	Very critical	Extremely critical
	1	2	3	4	5	1	2	3	4	5
8.1 Poor level of efficiency and quality of work										
8.2 Poor level of professionalism and entrepreneurship										
8.3 Resources shortages, especially in construction technology, management and finance										

9. **Land acquisition**	Degree of occurrence					Degree of influence				
	Not significant	Somewhat significant	Significant	Very significant	Extremely significant	Not critical	Somewhat critical	Critical	Very critical	Extremely critical
	1	2	3	4	5	1	2	3	4	5
9.1 Lack of a negotiating mechanism to make land acquisition compensation more market-oriented										
9.2 Bureaucracy in settling land acquisition disputes and claims										
9.3 Lack of a land acquisition compensation monitoring system										
9.4 Lack of clarity about valuation methods for compensation										
9.5 Lack of law enforcement to regulate the price of land acquired for infrastructure projects										

10. Institutional and legal weaknesses in infrastructure construction	Degree of occurrence					Degree of influence				
	Not significant	Somewhat significant	Significant	Very significant	Extremely significant	Not critical	Somewhat critical	Critical	Very critical	Extremely critical
	1	2	3	4	5	1	2	3	4	5
10.1 Obsolescence of building regulations										
10.2 Changing and inconsistent law and regulations										
10.3 Ineffectiveness of implementation of existing statutes and codes										
10.4 Bureaucracy in formal procedures relating to infrastructure project planning, construction permission and administration										

11. Please indicate any other related issues (other than those mentioned above) that could inhibit the efficiency of the funding invested in trade-related infrastructure construction for trade and economic growth in Vietnam. Please write your answer in the box below.

Thank you very much for your kind assistance. The findings of this study should go a long way to contributing towards economic development in Vietnam.

Appendix B
Interview Questions

1. According to several surveys conducted by international organizations, the quality of Vietnam's infrastructure has been identified as the biggest hindrance to the growth of the economy in general and the trade sector in particular. Would you please give me some thoughts on the current use of public investments in infrastructure to support trade and economic development in Vietnam?
2. According to a recent survey of major issues affecting the efficient use of public investments in infrastructure in Vietnam, a number of issues were identified from infrastructure planning and delivery, including: institutional weaknesses in planning; corruption in funded infrastructure projects; land acquisition problems; shortages of local firms' capabilities and resources; bureaucracy in infrastructure construction; transparency and accountability problems.

    a. From your experience, how are these problems affecting the efficient use of public investments in infrastructure in Vietnam?
    b. What kind of changes would you recommend to bring into the processes of infrastructure planning and delivery to improve the use of the public funding?

G. Dang and L. Sui Pheng, *Infrastructure Investments in Developing Economies*,
DOI 10.1007/978-981-287-248-7

Appendix C
Samples of Interview Records

Interview with G1

Question: "According to a recent survey of major issues affecting the efficient use of public investments in infrastructure in Vietnam, a number of issues were identified from infrastructure planning and delivery, including: institutional weaknesses in planning; corruption in funded infrastructure projects; land acquisition problems; shortages of local firms' capabilities and resources; bureaucracy in infrastructure construction; transparency and accountability problems. From your experience, would you please give me some thoughts on these issues and how are these problems affecting the efficient use of public investments in infrastructure in Vietnam?"

Answer: "Although it is called Ministry of Planning and Investments, more than 80 % of the MPI's work involves allocating investment budget. The MPI's development planning function only accounts for a small part. In other countries, development plans are made before allocating investment budget to various sectors and regions. In Vietnam, the ability to gather capital from state budget, loans and aids from international organizations would be taken into consideration before needs and interests of line ministries and local governments would be taken into account. As a result, the budget allocations never match the investment plans.

Development plans require a long term vision, e.g. for 5, 10 years. But, what planners mostly do is to allocate the investment budget on a yearly basis. For example, development plan to 2015 exists in paper only as there is no plan for budget allocation until 2015. Or for example, there are plans to build the road connecting Hanoi and Hai Phong, the National Assembly House, or some hydroelectric power plants (in 2015); but there is no confirmation whether there is enough capital for these projects in 2015. Meanwhile, capital available in 2012 would be allocated for other needs first. Thus, the five-year development plan to 2015 would never be delivered as planned. The plan would then be modified and budget allocation of Vietnam would never match the investment plans. Investment budget to be available in the coming fiscal year would be allocated to short-term needs arising at the time of decision making, rather than being linked to the long-term goals specified in development plans. Long-term development plans are prepared just for the sake of planning. Many investments were planned and

G. Dang and L. Sui Pheng, *Infrastructure Investments in Developing Economies*,
DOI 10.1007/978-981-287-248-7

approved by various government levels even though there is no corresponding plan of allocating capital investments.

The decision making by the National Assembly on whether the high-speed train connecting Hanoi and Ho-Chi-Minh City should be built is a good example. On the one hand, there are arguments that taking advantages of resources from Japan and international organizations, Vietnam should build the train, thus setting development direction of the national transportation system. Others argued that there are other short-term needs and thus resources should be allocated for these needs first, and the high-speed train could be built later. As a result, although public investments in Vietnam are long-term plans on the record, in reality they are short-term plans.

The next problem comes from the process of planning. In Vietnam, planning controls all investments and the MPI becomes the leading advisor of the central and local governments. However, lack of competent and qualified planners and the short-term vision resulted in such case as: for example, the construction of a 10 km road would take 5 years to be completed. This year, due to lack of capital and with short-term vision, planners would decide to build the road with 10.5 m width. But two years later, when there was heavy traffic on the road and after going abroad to visit some similar projects, they would think that a 10.5 m wide road was not enough, the road should be built with 30.5 m width. As a result, the road would be completed with some parts wider and some narrower.

In Hai Duong province, the construction of Ngo Quyen road is the same case. In Hanoi and other areas, construction of roads faces the same situation. Besides, due to the lack of state capital in the beginning, first parts of the roads would be built by the public. As a result, after the road is completed, the first parts of the roads would be narrow and the later parts would be wide. This is a typical problem in public investments in Vietnam. This affects the development of infrastructure and is difficult to be solved. Besides, due to lack of capital, roads may be completed without pavements and drainage systems. So, roads can be built without any plans for pavements, drainage systems, lighting systems and greenery. Land for these auxiliary structures thus would not be set aside. Instead, the locals would use the land along the roadsides for housing. So, when these auxiliary structures needed to be built after that, excavation on the site would be very costly and the execution would be very difficult.

The next problem is the lack of accountability. Many meetings would be held before a project proposal was approved. Attending these meetings were a number of representatives from various government agencies. But, no single government official had to take the responsibility for the decisions made. For instance, a number of committees would be set up to study project proposals. When a project was completed, a committee, whose chairman would have never been to the project site, was also responsible for the issuance of the official acceptance of work. Whether the quality of work is acceptable is not the responsibility of the chairman. For example, the issuance of the official acceptance of the Can Tho Bridge would require the approval of the MoC. Representing the MoC would be some doctors and professors whose have never been to the site. That is because the issuance is merely a formality. No government official is held accountable for their decisions and actions, which are intertwined with each others throughout the feasibility study, design appraisal and approval, and payment procedures. It is difficult to make the complicated and bureaucratic administrative system accountable. There are laws and regulations but there is a lack of law enforcement. This is one of major weaknesses that make public investments in Vietnam inefficient.

As a result, investments were scattered and incomplete, and things that must have been done first were done later, and vice versa. A good example is the case of the Thanh Tri Bridge. Because the design of Thanh Tri Bridge (the largest among 7 bridges across the Red River in Hanoi) was done within the transportation sector, it was quickly readied at the time of decision making. The construction of the bridge started in 2002. A lot of money thus had been invested in building it. However, the access roads to the bridge were not put on the table because of the difficulties in land acquisition, which involved sectors and the local government other than the transportation sector. As a result, after the bridge had been completed for more than 2 years, the access roads to the bridge were still under construction, and the bridge could not be opened for use.

And the last problem is land acquisition. The thinking behind land acquisition executed in Vietnam is not right. The resolution no. 5 has just been issued by the Central Committee (the Politburo) to canvass the public opinion to amend the Land Law. However, if we don't change our thinking we would never solve the issue entirely. Since the issuance of the Decree No. 22, the execution of land acquisition and compensation has been mixed together. But the execution of land compensation would involve negotiation with land users. In public projects, the negotiation would make land acquisition very difficult. Land acquisition and compensation should not have been mixed together. Land acquisition should be implemented first. And land compensation deals for related land users should be settled later. But in current procedures, land compensation is carried out first and followed by land acquisition. The decision to acquire land is only a formal decision because land could not be acquired (until land compensation deals are settled). The procedures have been mixed up. Land acquisition and compensation should be separated in which land should be acquired first for project development, and compensation deals would be settled later. The current procedures, in which compensation deals are settled first, make land acquisition more and more difficult. Difficulty in land acquisition is the main cause for time delay because if land could not be acquired, the construction work could not be started. So the execution of many infrastructure projects has been fragmented. Construction would start first in the area where the land acquisition was settled. Construction of other parts where the land areas have not been acquired would be delayed. As a result, it would be very difficult for project scheduling in Vietnam. For example, if you plan to complete the project on 31st December but you are not sure whether land acquisition would be completed on 31st December, how could you prepare the project schedule? As a result, project scheduling would be full of uncertainties.

So public investments would not be efficient with planning and scheduling full of uncertainties and lack of accountability among parties involved. It is necessary to revisit the public investments in Vietnam. There could be no growth if the current use of public investments is not improved".

Question: "So what kind of changes would you recommend?"

Answer: "Firstly, there should be changes in the institutional capacity and the methodology behind the decision making. Investment planning should be one-step ahead (of budget allocation). Investment planning (of MPI) should really focus on the planning function. Currently, investment planning and budget allocation functions are mixed together in Vietnam. However, planners only focus on the budget allocation, and ignore the planning function. Successful investments depend on good planning and resource allocation. You can have long-term or short-term plans, but capital resources have to be mobilized accordingly. Only after you have a good plan, you can start to mobilize capital resources. This can help match investment plans and budget allocation plans and the two types of plans can work together.

Secondly, the centrally produced plans are the formal documents setting directions and strategies to use public investments. These plans in reality are often ignored. The current use of public investments in fact is based on political leaders' will and serves in the interests of a small group of political leaders and government bureaucrats.

The next thing is the (functional and fiscal) decentralization associated with public investments in infrastructure. At lower levels, infrastructure investments require approval from local governments. At higher levels, infrastructure investments require approval from the National Assembly. But most members of the National Assembly actually have a limited knowledge of infrastructure investments. To make the right decisions, members of the National Assembly must have a reasonable level of competence in infrastructure investments. But among 500 members of the National Assembly, only a few are competent at infrastructure investments. The remaining 99 % without the required competence could not be able to make the right decisions.

The next problem is in the execution of infrastructure projects. There are three main parties involved in public funded infrastructure projects, including clients (developers), consultants (designers, engineers, supervision consultants, etc.) and contractors. The tripartite model makes it easier for the government to manage the state budget. However, the problem is most clients and developers of public investments are not competent at their jobs. For example, a principal of a school with 50 teachers was assigned to build a new school. Currently, the yearly operation cost of the school was about VND 1 million. However, the new project was a VND 50 million investment. He thus did not have the experience and capability to manage the 50 million dong investment. Despite the incompetence, he would still be appointed as the developer of the new development. Moreover, developers of public projects without the required competence would still want to undertake the jobs since this is the opportunity for them to abuse public funds. As a result, clients and developers of public projects often do not have the required capacity and level of professionalism.

Lacking the required capacity, clients in public projects employ consultants. In private projects, clients without the required competence would still have to employ consultants, but they would closely supervise the quality of their work. Meanwhile, in public projects, since clients do not have to spend their own money, they do not really care about the quality of their work. They would leave all the work to their consultants. The quality of concrete, grade of bricks, prices of materials, etc. all would be supervised by their consultants.

But, there are a lot of issues with local consultants. Although I am the director of the Construction Department of a province, I am not sure how many consultants are there in the province. There are too many consultants because local consultants just need to register for any consultancy services that they can think of, and then they would be licensed to provide the services. The next problem is that an architect can be an official employee of several consultants. He may work somewhere else but his name still exist in others' list of employees. In other cases, drawings issued by a consultant would be stamped by a professor in the Civil Engineering Department of the University through a contract, which may expire tomorrow. Or for example, a professor was responsible for preparing the drawings, but he would ask his fresh graduate students to do the job for him. That's the current state of local consultants. The same problems can be found in supervision consultants and project management consultants. It is very difficult to find an experienced and highly qualified consultant in the industry. As a result, the quality of consultancy services is very low. That is a common problem.

Next is about contractors. If you look at local contractors' bidding documents, you would think that they have the required human, equipment and financial resources. However, these resources appear on paper only. For example, a tower crane can be found in the list of equipment of a bidding firm, but its exact whereabouts was a mystery because the firm had no equipment rental contract. Or several leading experts were listed in the firm's project team members, but they would have never been seen on site. Contractors have very poor quality management systems. Most in-house quality management systems are removed because contractors themselves do not have long-term development plans, thus they do not want to spend their resources maintaining such a system. Even large firms hire informal labor forces. So the first problem with contractors is that their human resources have low level of professionalism. The second problem is that their in-house quality management is overlooked; they often depend on other parties including designers, supervision consultants and related government agencies for quality management. The last very important problem is about contractors' fraudulent practices, which affect the quality of infrastructure facilities, especially the underground structures. Since contractors often have to bribe clients and consultants to get the jobs, they want to compensate for the bribe expenses through material (concrete, steel, etc.) theft. If clients are really diligent in managing the projects and consultants are experienced, the projects can be effectively executed and the contractors can do their jobs effectively as well. But both clients and consultants have low level of professionalism and contractors also could not do their jobs effectively. Most contractors increase their profits through material theft.

Rarely do contractors focus on accelerating construction progress and saving construction costs. As a result, most projects are delayed and contractors would bribe government officials to cover their poor quality construction practices and outcomes. If contractors focus on accelerating construction progress, they can get their payments early, which ease all financial problems. For example, getting paid early, they don't have to get bank loans and pay interests, they can buy materials early on, and they can pay salaries and hire best workers, which in turn accelerate project progress. But, most contractors care about how much they can spend on the projects rather than about how they can deliver the projects effectively. It is really a wasteful way.

All these problems affect the quality of public projects from feasibility study, design appraisal and approval to execution. Even in projects that involved some form of partnership with the private sector such as exchange private land development approval for financing of public infrastructure, ultimately these problems would waste national resources. Investment decisions are made in the interests of a small group of people. The laws on canvassing the public opinion are under consideration. However, the actual decision making process is complex and opaque to the general public. In fact, decisions are made by authorized government agencies regardless the public opinion. For example, an infrastructure investment would have been approved by the Prime Minister, but the government would still canvass the public opinion for show. Since the public knows that they have no say in the decision, the public also shows little interest in the decision making. The public only cares about their private interests. Even political leaders and government bureaucrats pursue their self-interests as oppose to national interests. As a result, national resources are wasted in a lot of public investments. The pursuit of personal, sectoral or regional interests of politicians and government bureaucrats and the lack of coordination result in excessively spread investments in seaports, airports, industrial zones and urban areas. Meanwhile, budget allocation is scattered. Every year, state budget would be allocated to every region from Hanoi, Ho-Chi-Minh City, Hai Phong, etc., rather than being gathered to complete some key projects. This is one of the major issues that make infrastructure development incomplete and delayed".

Question: "As far as I know, the relevant authorities are assigned as clients or developers of public investments in infrastructure?"

Answer: "Yes, that's right. For example, the role of a developer of a national high way would be assigned to the Ministry of Transport. For example, Project Management Unit Number X is one of the representatives of the MoT in the development of some transport infrastructure projects. But, they only do three things: first, spending public funds allocated; second, selecting consultants and contractors (bidding is just for show, they can manipulate the process to select firms they want); and third, making decisions when and how to develop the infrastructure. They leave all the remaining jobs to their consultants. And they are not held any liability for the project performance".

Question: "So what would you recommend to change the situation?"

Answer: "Several changes can be made. First, when being assigned as clients or developers of a public investment in infrastructure, the developers or PMUs should have the required capability and a clear action plan. PMUs have to be able to make strategic decisions about the project performance and they have to be accountable for their actions. If PMUs have the capability to manage project effectively, the project quality and progress would be improved.

Since the current mechanism does not tie any specific liability to the performance of the assigned clients and developers, they fulfil their tasks with poor diligence. Moreover, selected developers might not be those who have the professional competence, but those who have the same personal interests. The poorly selected developers only focus on abusing the state budget allocated. For example, in order to get more money to spend, they might approve a construction method whose cost would be three times more than material and manpower costs".

Question: "So, there are issues with both government agencies and local firms?"

Answer: "That's right. You should combine issues from both sides. First, that's right. There are a lot of issues with local firms. Because the current mechanism lacks transparency and accountability, firms often exploit tax loopholes and involve fraudulent practices to increase their profits. For example, the standard concrete thickness of the underground structure of a road should be 25 cm, the actual concrete thickness can be only 20 cm. Firms would then bribe related authorities to cover the poor quality. Since there is a lack of accountability in current infrastructure development, firms can maximize their profits by lowering the quality standards.

Besides, the lack of competent and qualified government officials constrains the government's ability to formulate development strategies and set clear laws and regulations. Lawmakers copied laws and regulations from different countries without considering specific conditions of Vietnam. Laws and regulations are consistently changed and inconsistent. Moreover, the slow issuance of circulars describing the duties of relevant parties to comply with regulations would also diminish their effectiveness. For example, increased salaries for workers came into effect since the first of July. But there was no guidance about how the increased salaries would be claimed until December (in public projects, workers can claim the increased salaries). A lot of workers would already leave their employers by December. As a result, the increased salaries claimed after December would go into the employers' pocket rather than the workers.

So, it all comes down to the institutional capacity of the government, including the law, the planning process and the vision of top leaders. The planning process should be more transparent. Development plans should be made with clear objectives and priorities. Once the planning and decision making were carefully done, the implementation of the development plans would be more efficient".

Interview with G4

Question: "According to several surveys conducted by international organizations, the quality of Vietnam's infrastructure has been identified as the biggest hindrance to the growth of the economy in general and the trade sector in particular. Would you please give me some thoughts on the current use of public investments in infrastructure to support trade and economic development in Vietnam?"

Answer: "Since infrastructure is the backbone of the country's economy, the government of Vietnam has put a lot of effort into infrastructure development over the years. Although there was some growth, the economy has been held back by the inadequacies of the infrastructure system. Recognizing the issues, the government has taken action to deal with the issues. In order to accelerate infrastructure development, the government has been getting loans to finance infrastructure investments. Over the years, loans from international organizations such as IMF, WB or ADB or from donor countries through bilateral assistance in the form of ODA have been mainly spent on infrastructure development. Annually, investments in infrastructure account for about 40 % of total public investments. Large loans are mainly used to finance infrastructure investments. Major projects around the country are all in the infrastructure sector. Among them, the first major project was the national high way No. 1, the most important route connecting North and South Vietnam. The project was funded by loans from WB. After that there have been a lot of roads and bridges financed by such loans. Contributions of these projects in the development have been considerable. Using loans from international organizations, Vietnam has to follow international rules. Using these loans thus has brought about two important changes/benefits in our country. Firstly, new technologies such as new technologies in building bridges or tunnels that we could never dream of have been transferred to Vietnam. For example, the building technology of the My Thuan Bridge, the first stayed-cable bridge, has been transferred to Vietnam with the support from Australia. We had studied the technology for years, but we were never successful. With technology transfer, we have been able to use the technology by ourselves. For example, the building technologies for the Thanh Tri Bridge or the Song Gianh Bridge are very difficult. Since the 1960s, Vietnam had studied these technologies. But the largest span we could build was only 60 m. Now, we can build bridges with 120–135 m spans. So besides providing the physical facilities for the economy, infrastructure development has brought about an improvement in building capacity.

The second benefit of using foreign loans for these projects is that we have also been able to improve our management capability and professionalism. For example, many procurement procedures and regulations have been improved under the direction of the international organizations such as criteria for assessing design, construction methods, monitoring, auditing, etc., which in turn improve the transparency and accountability in the sector.

With several decades of work experience in public investments in infrastructure, I think that these are the two important changes. Besides, infrastructure development has been able to support our economic growth. For example, previously it would take us half a day to go to Quang Ninh. Now it only takes us 2 hours. Newly built airports help integrate the national market and connecting it to markets in other countries. The upgraded seaport system improves our capacity of sea transport for international trade.

So, achievements of public investments in infrastructure have been considerable. However, there have been issues with the efficiency in public investments in infrastructure. Since there have been high demand for infrastructure development during periods of rapid growth, we built too many seaports, airports, etc. As a result, investments have been spread thinly across these projects, thus compromising their efficiency. For example, local governments of every province want to have airports. Seaside provinces want to have seaports. Thus, investment planning in Vietnam lacks a long-term vision. For example, the newly built Can Tho International Airport was expected to open for international flights. However, we have not prepared any plans on how to attract international flights to the airport. So the problem with public investments in infrastructure in Vietnam is that although these investments were able to support economic growth, the excessively thinly spread investments have reduced their efficiency.

The second problem is the lack of coordination in investments. Some large seaports have been built without any passage leading to the ports. For example, there is a very large seaport in Ho Chi Minh City, which was expected to accommodate large ships. However, ships larger than 15,000 DWT could not go into the seaport because the passage leading to the port is very shallow. Some bridges have been built without access roads to the bridges. For example, the newly built Phu My Bridge in Ho Chi Minh City could not connect District No. 2 and District No. 7 because the construction of access roads has been delayed by land acquisition issues.

The third problem is about the poor quality of infrastructure facilities. For example, the Phap Van-Cau Gie Road was expected to be a first class road. However, the quality of the newly built road is only enough to be qualified for a third class road. So, quality is an important issue. The causal factors of the poor quality can be found in various stages from soil investigation, design, and construction. The most recent quality incident was the settlement in the Sai Gon-Trung Luong highway. The weak soil foundation had not been handled very well, thus making the project inefficient. So quality is the third problem.

The fourth problem is the lack of coordination in land acquisition. Many projects have been delayed because time delay has not been considered an important issue. Time is an important factor behind the investment efficiency. For example, investment efficiency of a road can be improved significantly by completing the road just one day earlier. But the issue seems to be unnoticeable in Vietnam. Delay in land acquisition is an important factor causing the inefficiency in Vietnam. The lack of coordination between centrally planned investments and land use policy at the local government level is the major cause of the delay. For example, in a 100 km road project, construction of only the first 80 km can be executed; construction of the remaining 20 km could not be started because land acquisition for the area had not been settled. A lot of projects could not be completed in time because of the same issue.

Besides, if there is one more setback in infrastructure development in Vietnam, it is the rising cost of infrastructure development compared to other countries. For example, development cost for roads located whether in urban areas or not is very high. There are two causes of the rising cost. Firstly, the cost can be naturally high because of the project complexities. For example, dealing with the weak soil foundation of the Sai Gon-Trung Luong highway could cost from USD 5–10 million per km, which is very high compared to that in other countries. But the cost can be artificially high. For example, several modern roads in Hanoi are considered the most expensive roads built in the world. Construction of these roads may cost half the cost for land acquisition. For example, construction cost for the transport infrastructure at Kim Lien (Hanoi), the key loop in the city, was about VND 21 million, but land acquisition cost more than VND 80 million. These serve to push up the price artificially".

Question: "Yes, you are right. These issues can have major effects on the efficiency of public investments in infrastructure in Vietnam. You mentioned that the thinly spread and uncoordinated investments is one of the major issues. What kind of changes would you recommend to improve the quality of development planning?"

Answer: "It is important to review the master plan for infrastructure development, e.g. master plan for seaports and airports. Of course, there should be a leading agency to review development plans and strategies. Investments are scattered because local governments have their own master plans for building airports and seaports in the local area to support local economic growth, securing political positions or competing for scarce funds, etc. The problem requires the government to re-examine the development plans (for seaports, airports and roads) and make necessary adjustment to improve the efficiency".

Question: "As far as I know, the national development plans (both short-term and long-term) are prepared based on balancing the needs and interests of different sectors and local governments. Why do you think there is a lack of coordination in planning?"

Answer: "That's right. Planning is based on the needs and interests of line ministries and local governments. However, planners lack a long-term vision of developing the economy as a whole. For example, every province wants to have a seaport. However, local needs are not coordinated and prioritized to achieve national goals. As a result, plans adopted by individual regions can be unnecessarily duplicated and contradictory. Besides, with many needs, capital resources have to be allocated too thinly for many projects. The lack of capital in turn delays project progress".

Question: "Earlier, you mentioned the quality issues of infrastructure construction. In your opinion, how can the quality be improved?"

Answer: "The poor quality of infrastructure construction can be found in several aspects. Firstly, there is a lack of professional capacity for design coordination, review and checks. For example, the topography surveys are not sufficiently carried out to provide accurate and detailed data for design. During the construction phases, the data are often not checked by additional topography surveys as well. However, budget for topographical surveys is often limited. It is safe to say that 60 % of quality incidents are caused by inadequate data from topographical surveys. The current government regulation of topographical surveys requires that additional surveys should be carried out in the design or construction phases if there is any doubt about the data provided in previous phases. So besides inadequate professional capacity, financial issue is also a cause of the poor quality in infrastructure construction.

In addition, inadequate supervision also affects the quality of infrastructure construction. Workmanship, material quality and construction methods all require close supervision. However, supervision in Vietnam hasn't been done professionally. Even when foreign supervision consultants were employed, they would hire local engineers who did not have the required capability. Capacity of local supervision consultants is one of the major concerns. If close supervision was carried out, the project efficiency would be very high. For example, in the construction of the Sai Gon–Trung Luong road, asphalt coating requires supervision of the standard temperature of the asphalt when being poured and standard conditions for coating. If it is raining, it would not be recommended to proceed with the coating. However, if the supervision was poor, asphalt coating would still proceed. So, developers' supervision consultants play an important role in preventing contractors from lowering quality standards".

Question: "So what would you recommend to improve the supervision?"

Answer: "There are two aspects. Firstly, supervision staff should be trained professionally. Foreign developers often hire foreign supervision consultants with international prestige. The quality of these projects is very high. Meanwhile, local supervision consultants hired by local developers often lack the required professional and ethical capability. It is common in Vietnam that local supervision consultants accept bribes from contractors to cover their poor quality works. The collusion between supervision consultants and contractors to overlook poor quality materials and works is one of the major issues affecting project quality. Although local supervision consultants were able to provide supervision staff in quantity, they lack highly qualified and ethical professionals in their workforce. It has been difficult to mobilize and retain skilled staff in the supervision consultancy profession. Low income is considered the major cause of the issue. Despite its important role in quality assurance, the supervision consultancy profession has not received adequate attention in the industry".

Question: "You also mentioned that land acquisition is one of major factors affecting the quality of infrastructure construction. What would you recommend to address the issue and to accelerate land acquisition?"

Answer: "In order to have an effective land acquisition, the government should prepare a clear spatial plan for infrastructure development, based on which land would be set aside in advance for future infrastructure development. This way we can manage land resources for infrastructure development effectively, and land resources for infrastructure would not be disturbed by other projects. Secondly, in the case where infrastructure development would have to involve in resettlement, we need to do two things.

Firstly, because of the public purpose of national infrastructure projects, the government should promote citizens' obligations to serve national interests. Secondly, the government should provide adequate compensation packages for displaced land users. For example, the government regulations indicate that displaced land users would be relocated to resettlement sites with conditions at least equal to the previous one. But there is no mechanism to ensure that adequate compensation is provided. Plans for transition to a new way of life should be prepared as well. For example, living in a 6 m^2 house in central Hanoi, a family's business is so good that they have been able to afford to send their kids to study abroad. But, if they were relocated to a 60 m^2 apartment, which is 10 times bigger than the current one, but located far from the business area, they would lose their business and their income. It is clear that the government should have a policy to maintain the livelihood for displaced land users. The current mechanism has recently created much dissatisfaction among displaced land users.

Finally, land acquisition should be done consistently over entire sites. Currently, the acquisition of a new site can be inconsistent. The acquisition of the area with compliant land users can be completed quickly. But the acquisition of the remaining areas can be delayed. The inconsistent acquisition can cause social inequality among displaced land users, and affect the efficiency of infrastructure development. Providing readily available sites for infrastructure development is thus important.

Some local governments have been able to solve land acquisition issues. For example, the local government of Da Nang has been able to provide readily available sites for infrastructure development. When there are infrastructure improvements, land value in surrounding areas often increases. The local government of Da Nang has taken an initiative to capture the increased land value to accelerate land acquisition. For example, if a road was planned to be widened by 50 m, the local government would acquire not only the 50 m but also another 50 m along the site, which would then be sold to compensate the sitting tenants and finance the road improvement project. Everyone would thus benefit from the revenue, creating social equality. However, the method might not be applicable to Hanoi where the acquisition would involve houses of many political leaders".

Question: "Are there any other factors affecting the quality of infrastructure construction?"

Answer: "About the resources of local firms, most firms involving infrastructure projects have financial difficulties. Secondly, the poor quality of infrastructure construction is also a result of the weaknesses in laws and regulations, especially the Tendering Law. A number of selection criteria specified by the Tendering Law are qualitative. The only quantitative criterion is the tender price. As a result, firms with the lowest tender price are often selected. Many Chinese firms took advantage of the regulation to get their jobs in Vietnam. But as I often said, the cheapness and top quality of projects could never come together. Projects executed by Chinese firms are often delayed. In many cases, Chinese firms could not complete their contracts and have abandoned their projects.

So, the most important factor affecting the efficiency is development planning, and then followed by other factors I mentioned earlier".

MIX
Papier aus verantwortungsvollen Quellen
Paper from responsible sources
FSC® C105338

If you have any concerns about our products,
you can contact us on
ProductSafety@springernature.com

In case Publisher is established outside the EU,
the EU authorized representative is:
Springer Nature Customer Service Center GmbH
Europaplatz 3, 69115 Heidelberg, Germany

Printed by Libri Plureos GmbH
in Hamburg, Germany